MW01505343

FOUNDATIONAL ENGINEERING GRAPHICS: PRINCIPLES AND APPLICATIONS

Ernst - Kelly - Clark - Glimcher

Published by RAIDER Press

Foundational Engineering Graphics: Principles and Applications

Dr. Jeremy V. Ernst, Dr. Daniel P. Kelly, Dr. Aaron C. Clark, Dr. Shelley A. Glimcher

Copyright © 2021 by Dr. Jeremy V. Ernst, Dr. Daniel P. Kelly, Dr. Aaron C. Clark, & Dr. Shelley A. Glimcher

ISBN 978-1-7368376-0-3 (paperback)
ISBN 978-1-7368376-1-0 (open access ebook)

Published by RAIDER Press

Distributed by RAIDER Press

Edited by:
- Dr. Deidre L. Kelly - Indiana University
- Erik J. Schettig - North Carolina State University

Contributors:
- Dr. Kevin G. Sutton - North Carolina State University
- Dr. Theodore J. Branoff - Illinois State University
- Dr. Joshua W. Brown - Illinois State University
- J. Parks Newby - North Carolina State University
- Engineering Graphics instructors and students for providing feedback

All images not credited were developed by authors and contributors of Foundational Engineering Graphics: Principles and Applications

Author Bios

Dr. Jeremy Ernst, DTE is Professor of Technology and Associate Chancellor for Research within the Worldwide Campus at Embry-Riddle Aeronautical University. He has had prior academic and administrative appointments at Virginia Tech as well as North Carolina State University. His efforts center on curriculum research and development in STEM education to provide evidence-based models that promote engagement, development of cognitive competency sets, and performance-based application abilities of students at-risk.

Dr. Daniel P. Kelly is an Assistant Professor of STEM Education at Texas Tech University in the Department of Curriculum and Instruction. He earned his doctorate in Technology Education from North Carolina State University where he also served on the faculty. Previously, he worked as a middle and high school science, technology, and engineering teacher in North Carolina. Dr. Kelly serves as the Associate Editor of the Engineering Design Graphics Journal and Editor-in-Chief and Founder of the Journal of Foster Care. Dr. Kelly studies how STEM education and engagement can improve the educational outcomes of students at risk of not completing high school due to academic, behavioral, or social needs. Of particular interest are children in foster care and other non-parental custody arrangements.

Dr. Aaron C. Clark, DTE is Department Head and Professor for Science, Technology, Engineering and Mathematics Education within the College of Education at North Carolina State University. He is a member of the Technology, Engineering and Design Education faculty. Research areas include graphics education, engineering education, visual science and professional development. He has also served in various leadership roles in disciplines related to engineering education and career and technical education. Dr. Clark is recognized as a Distinguished Technology Educator by the International Technology Engineering Education Association and for the American Society of Engineering Education; Engineering Design Graphics Division.

Dr. Shelley Glimcher is a recent graduate from the Technology, Engineering, and Design Education Program at North Carolina State University. Previously, she worked extensively with the First-Year Engineering Honors Program at The Ohio State University and continues to work with the program in an advisory role. Her teaching interests focus on first-year cornerstone engineering design experiences and technical graphics with a complimentary research interest in tactile spatial thinking.

Any opinions, findings, and conclusions or recommendations expressed in this material are those of the author(s) and do not necessarily reflect the views of the National Science Foundation.

This material is based upon work supported by the National Science Foundation under Grant No. 1900348

Table of Contents

This page intentionally left blank

UNIT 1

SKETCHING

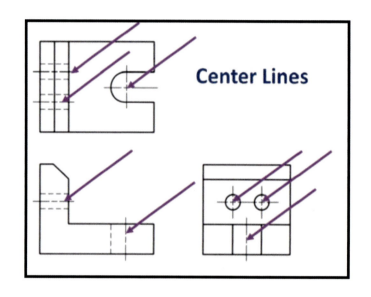

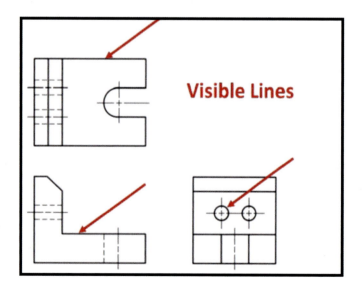

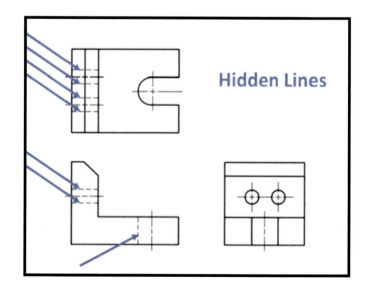

Topic Introduction

Sketching is a critical skill and is used extensively at all levels of the cyclical engineering design process with various levels of complexity. You may be surprised to learn that the overwhelming majority of communication in the engineering design process has a graphical component to help explain the object of interest. Sketches range from very rough hand drawings to just an outline conveying a general thought. They can also be very detailed illustrations to help communicate a complex design.

Chances are you have had practice with sketching in art class or technology classes in the past. Now we will look at sketching in the context of technical communication. This unit will cover different types of sketches, various sketching techniques and their uses, and give you opportunities to practice these skills. As seen in figure 1.1 of the Space Needle, hand sketches combined with text can used to communicate preliminary designs to a wide audience.

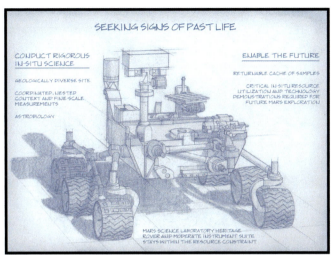

Figure 1.2 Mars Curiosity Rover - Image courtesy of NASA

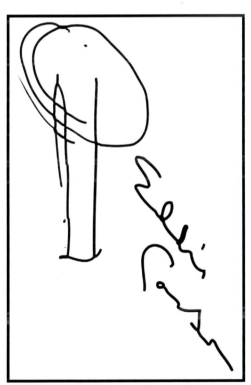

Figure 1.1 Space Needle - Image courtesy of Space Needle History

Everyday Uses

Sketching has many uses in the design cycle and gives engineers and designers a method to quickly convey information to others. For starters, ideas are often brainstormed using rough design sketches, objects are typically manufactured using specifications from technical drawings, and then illustrations are frequently used to communicate the overall design to a wide variety of people (see figure 1.2). Whether you realize it or not, you have probably seen each type of these drawings in everyday life. Let's go through some examples that highlight the use of a rough design sketch, technical sketch, and illustration with familiar objects.

Did you know that the concept of the Seattle Space Needle was sketched on a napkin in the 1950s (see figure 1.1)? Rough sketches have just enough information to convey an idea but are often lacking the technical details, such as dimensions or assembly instructions, to create the object itself.

Figure 1.3 is an example of a technical multiview projection sketch for the back of a GoPro case. This multiview sketch clearly shows the overall shape and scale of the different features with appropriate line-types for visible, hidden, and curved lines. While there is not enough size information for the object to be actually manufactured, the idea of the part is clearly communicated and could be used to help develop a prototype. The idea of adding size information is covered later in the dimensioning unit.

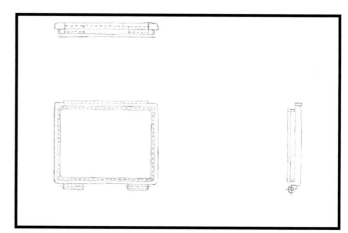

Figure 1.3 - GoPro Case

Figure 1.4 shows an artist's conceptual rendering of NASA's Skylab Orbital Workshop, the precursor to the International Space Station we know today. This type of visual communication document is often found in advertising. It has the most visual detail without overwhelmingly technical details found in dimensioned detail drawings created for manufacturing.

The Basics

There are many different ways to visually communicate a design idea to a wide audience. Sometimes a rough sketch will suffice; other times more detail is needed. Learning technical graphics will help you develop this valuable communication skill set. In this unit, we will explore different types of sketching techniques involving outlines, identifying and using different line types, sketching techniques for lines and arcs, and scale. Learning how to sketch will give you a solid foundation for more advanced topics. Understanding certain nuances or rules of sketching, like the use of hidden lines, will help you interpret more complex drawings.

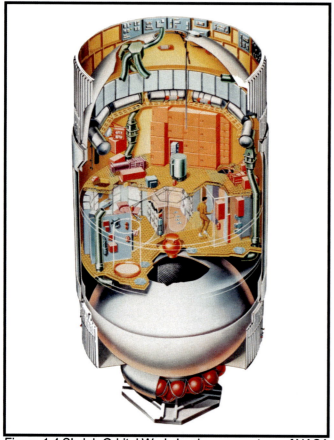

Figure 1.4 Skylab Orbital WorkshopImage courtesy of NASA

Figure 1.5 is an example of a computer aided design (CAD) assembly illustration of a turkey injector.

Figure 1.6 is a multiview projection sketch of just the cylinder piece of the turkey injector from figure 1.6. The different line types are labeled in the sketch. Notice how the front view and the right side view are aligned with each other.

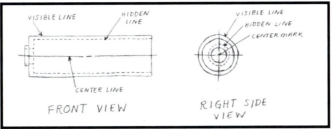

Figure 1.6 - Multiview Projection of Turkey Baster Cylinder

Contour Sketches

One way to show the general features of an object is to create a contour sketch. This technique involves simply showing the outline of an object. Figures 1.9 and 1.10 are some examples of some familiar objects shown in a contour sketch. Note these sketches are not perfect, but still clearly communicate the main shape and proportions of the objects.

Figure 1.8 - Hammer

Figure 1.7 - Roman Helmet

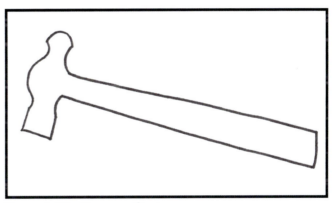

Figure 1.10 - Hammer Contour

Figure 1.9 - Roman Helmet Contour

The contour sketching technique can also be used with more complicated objects. For example, a Roman helmet (figure 1.7) is a more complex object but even a basic contour sketch can convey the general shape. Remember, contour sketches are intended to be a rough draft that communicate the basic shape and proportion of the object rather than all the minor, non-essential details.

Revolves

Learning how to make contour sketches is useful because outline sketches are necessary to make revolved objects in CAD programs. A revolved object is created by rotating a closed outline around a defined axis or outline edge. Let's look at the examples in figures 1.11, 1.12, and 1.13.

Figure 1.11 - Plant Stand

Figure 1.12 - Flower Pot

Figure 1.13 - Vase

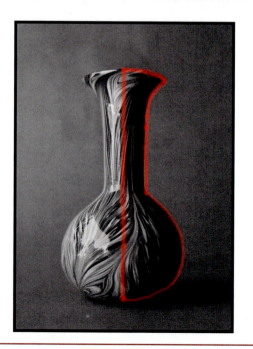

Figures 1.14, 1.15, and 1.16 demonstrate how these shapes would be created in a three-dimensional (3-D) solid modeling program using the revolve feature.

Figure 1.14 - Plant Stand

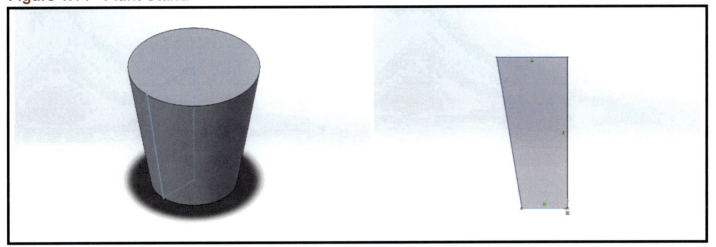

Figure 1.15 - Flower Pot

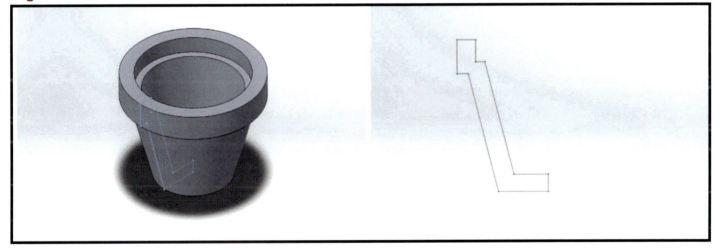

Figure 1.16 - Vase

Line Types

Sketching is an essential skill in the brainstorming portion of an engineering design cycle. Many designers keep a notebook handy so they can quickly sketch a wide variety of ideas before selecting concepts to investigate further. These rough sketches are then translated into technical drawings or illustrations to clearly communicate the idea to others and help with making prototypes.

There are multiple ways to go about sketching any object. It is important to learn some basic skills, such as how to sketch a straight line or arc, that should be practiced before trying to draw more complex objects. Figure 1.17 shows examples of the basic line types.

Basic Line Types

Construction (thin and light) ————————————

Visible (thick and dark) ————————————

Hidden (thin and dark) – – – – – – – – – – –

Center (thin and dark) ————— — —————

Figure 1.17 - Basic Line Types

- Construction Lines - Drawn first while sketching to show basic shape of object, edges, and features. Later, construction lines will be replaced with visible lines.
- Visible Lines - Show details of object with darker, more visible lines.
- Hidden Lines - Show features hidden by visible surface.
- Center Line - Indicate when a circular shape, such as an arc or hole, is present.

Center Mark and Center Line

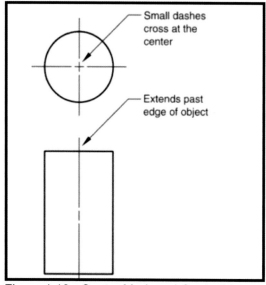

Small dashes cross at the center

Extends past edge of object

- Top Circle - Dashes show center mark.
- Bottom Image - Sideview of cylinder identifies center of hole.
- Both - Center lines extend past edge of object.

Figure 1.18 - Center Mark and Center Line

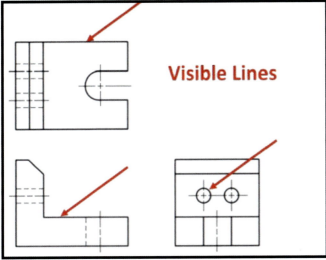
Figure 1.19 - Visible Lines

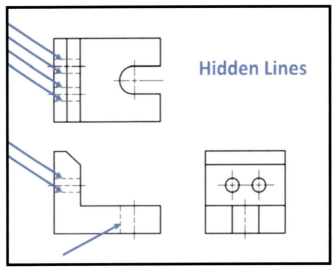
Figure 1.20 - Hidden Lines

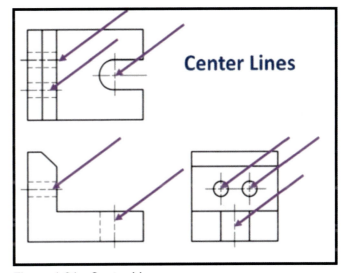
Figure 1.21 - Center Lines

- Visible lines (figure 1.19) show the features of the object.
- Hidden lines (figure 1.20) show the feature hidden by visible surfaces.
- Notice center lines and center marks (figure 1.21) extend past the visible lines.

There are many techniques sketch straight lines by hand, and it takes practice to find the method that works best for you. For starters, we suggest lightly sketching a straight line, and then drawing a darker line over the light sketch.

When you sketch something, it is important to be comfortable and in a setting where you can move around. Use a plain white piece of paper to practice sketching. It is important to draw towards you and to have a start and finish point when sketching lines. To maintain drawing straight lines, move your hand along with your arm instead of just moving your hand and wrist. This will prevent curves in your line and help.

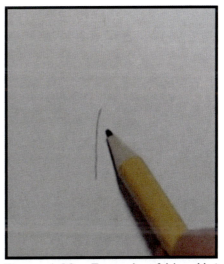
Figure 1.22 - Example of Line Not Drawn Straight

Try these methods to practice drawing lines (see figure 1.23):

- Draw out your two points, and then connect the dots from point A to point B.
- Draw out your line lightly (a construction line). Once you have your line, sketch over it to darken in the line.
- Lightly draw a dotted line and then connect the points to draw a visible line.
- Lightly move your pencil back and forth to lightly draw your line and then darken it in the draw a visible line.

While these are all different methods, it is important to have a comfortable method of drawing lines so that lines are straight and visible.

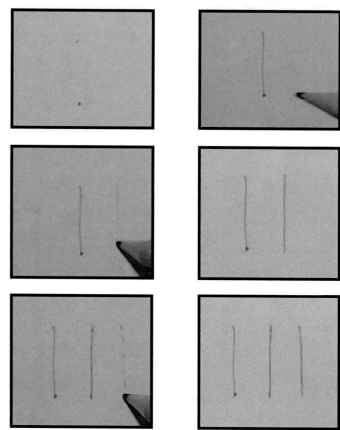

Figure 1.23 - Sketching Lines Methods

Sketching a circle can be tricky. We have a five-step process you can use to help construct better circles and arcs when sketching. Let's say we want to make a circle that is 2 inches in diameter. Here is the step-by-step method:

1. Make a square with the sides the same length as the circle's diameter.
2. Put crosshairs along the diagonals in the box.
3. Put hash marks on the midpoints of the sides and then about two-thirds of the way up on the crosshairs.
4. Connect the hash marks with small arcs until you complete the circle.
5. Add a centermark and radiating centerlines (optional: erase construction lines).

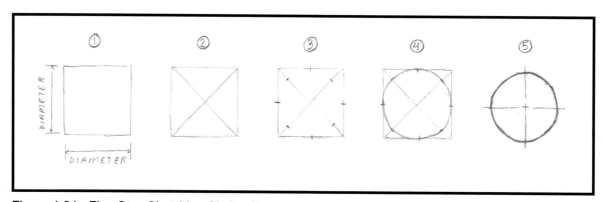

Figure 1.24 - Five-Step Sketching Circles Process

This technique can be used to make circles and arcs of various sizes. Practice making circles on a spare sheet of paper using this method.

Foundational Engineering Graphics: Principles and Applications

Paper Types

While a wide variety of paper exists that can be used for sketching, technical drawings typically rely on the use of regular grid and/or isometric grid paper. The regular grid paper is for 2-D multiview drawings, whereas the isometric grid paper is primarily used for 3-D isometric drawings (see figure 1.25) We will use both types of paper for exercises in this unit. If you do not have this type of paper available, you can print a PDF for regular grid paper and isometric paper in Appendix B.

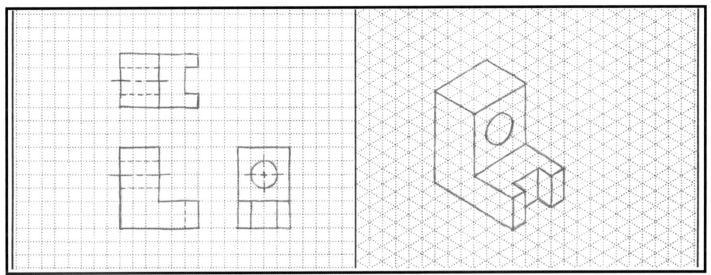

Figure 1.25 - Multiview Drawing

The sketching methodology is a step-by-step procedure:
1. Outline the height, width, and depth of the object to create bounds for the drawing.
2. Block in the main shapes of the object.
3. Sketch arcs/circles, and add hidden lines.
4. Darken the visible lines, add necessary annotations, and clean up construction lines.

You can use this sketching process even for objects you are unfamiliar with but may see during in-class or homework assignments for technical graphics.

Now let's practice some sketching using a flange mount (see figure 1.26). This part helps connect two objects and can come in many materials, shapes, and sizes. In this example, we will sketch only the front view together, but will provide self-check answers for the top and right side view for extra practice. Again, the most important thing to consider is showing correct proportions.

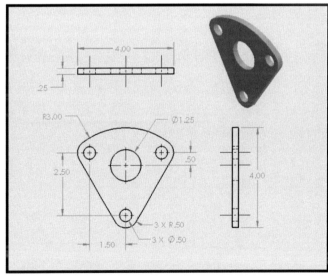

Figure 1.26 - Flange Multiview Drawing

There are multiple ways to approach this sketch, here are some steps to help you with the process.

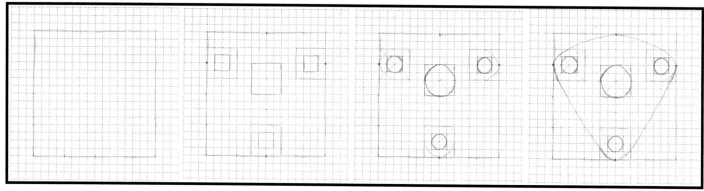

Figure 1.27 -Flange Sketch

Your finished sketch should look similar figure 1.28.

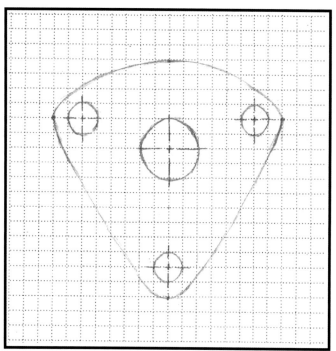

Figure 1.28 -Flange Final Sketch

Think about an aluminum can used for carbonated beverages. Regardless of manufacturer, the cans have the same general shape and operate in a similar manner. Using regular grid paper, draw the top and the front view of an aluminum can omitting the tab and can thickness. Your drawing should resemble figure 1.29.

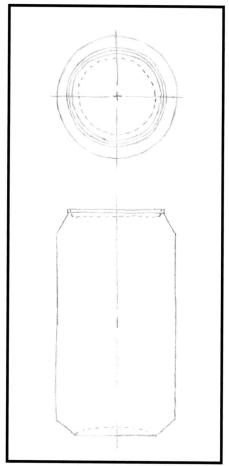

Figure 1.29 -Can Sketch

Scale

The concept of scale is necessary when communicating objects that are much larger or much smaller than a standard sized sheet of paper. In the title block of a technical drawing, you will notice that there is a space designated to specify the scale. The designation of 1:1 means that the object is the same size in real life as it is shown on the paper. Nesting dolls in figure 1.30 demonstrate how the same figure appears at different scales.

Often, objects are much larger than the sheet of paper, in fact they may be much larger than a person as seen in the telescope mirror comparison in figure 1.31.

Figure 1.30 -Nesting Dolls

The Hubble Mirror is approximately 94.5" (2.4 meters). So, we can scale the Hubble Mirror to a standard A** size sheet of paper (8.5" x 11"), by using a scale of 1:12 (Drawing:Object). That means that the drawing is 1/12 of the actual size of the object so the mirror would be drawn at approximately 7.9 inches on the standard sheet of paper.

Figure 1.31 - Telescope Mirror - Image courtesy of NASA

On the contrary, an object much smaller than a standard size sheet of paper will need to be shown much larger than in real life in a detailed drawing. For example, a gear in a mechanical watch (see figure 1.32) is an object you can see but because of its detail, it would be difficult to clearly show the shape and annotations (dimensions) adequately without drawing the gear larger than true size. So if this gear was 0.40" (~1 cm), it may be shown clearly on a sheet of standard sized sheet of paper (8.5" by 11") using a scale of 10:1 (Drawing:Object). That means that the drawing of the gear is 10 times larger than the actual size of the gear and would appear as a 4" object on the drawing.

Figure 1.32 - Watch Gear

Reflection Questions

1. Think back to a project you have done in the past where you used a rough sketch to communicate in some way to others. What drawings did you create and how did the drawings help you in your project?

2. What object is represented by the contour sketch in figure 1.33? How do you know?

Figure 1.33 - Contour Image

Activity

Contour Sketching

Materials Needed: Paper and Pencil

Let's practice some contour sketching. Here is a general everyday object for you to try:
Use a pencil to sketch the contour of the cup in figure 1.34. It should look similar to the sketch of the scissors in figure 1.33.

Figure 1.34 - Cup

Revolve Sketching

Materials Needed: Paper and Pencil

Let's practice some revolve sketching. Here is a general everyday object for you to try:
Use a pencil to sketch the shape of the bowl in figure 1.35 that would be needed to revolve around a center axis to recreate the shape. It should look similar to the revolved sketches on the previous pages. Your revolve contour sketch of the bowl should look similar to figure 1.36.

Figure 1.35 - Bowl

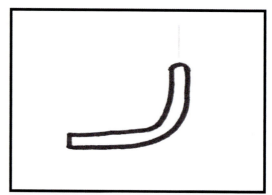
Figure 1.36 - Bowl contour

Quiz

 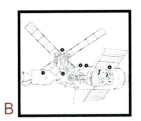 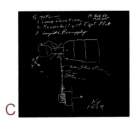

A   B   C

Images courtesy of NASA

For each of the following questions, choose A, B, or C for which image matches the description.

1. Which picture shows a rough sketch of a new product?
2. Which picture shows a technical sketch of a prototype?
3. Which picture shows an illustration of a concept for a new product?
4. Which picture would be presented to peers for feedback?
5. Which picture would be presented to a technical audience to discuss the project?
6. Which picture shows would be presented to a non-technical audience (i.e. a sales pitch)?

7. Which objects in the figures below would you need to scale down (make smaller), scale up (make larger), or be able to show as is on a single piece of 8.5"x11" sheet of paper?

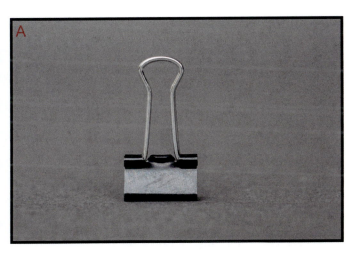 

Solutions in Appendix

UNIT 2

ENGINEERING GEOMETRY

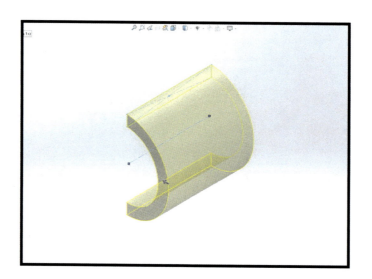

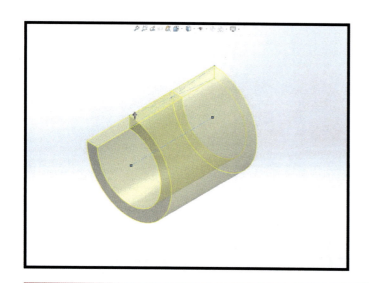

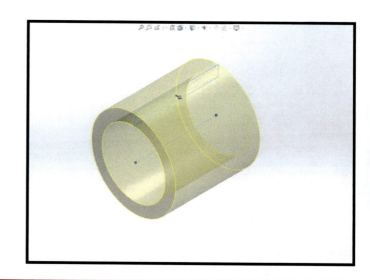

Unit 2 - Engineering Geometry

Introduction

Engineering geometry is an important part of engineering graphics. It involves understanding classification of geometric forms (point, line, surfaces, solids) as well as a wide variety of geometric relations (i.e. parallel, perpendicular, concentric). Understanding these underlying principles and being able to combine various geometric elements is what allows for modeling increasingly complex designs. Mastering material in this unit will help you create CAD-based objects in a variety of 3-D modeling software.

Think back to when you may have played with wooden blocks or Lego bricks to make a wide variety of creations, perhaps like the camera in figure 2.1. Basic shapes combined together can create complex objects. One topic you will learn about in this unit is identifying these basic shapes so you can break

down and model complex objects.

In one way or another, you have been applying the principles of engineering geometry for many years. Now it will be presented from the viewpoint of technical graphics. In this unit, we will review different types of coordinate systems, geometric relationships, and modeling items based on Boolean operations.

Figure 2.1 - Lego Camera Model

Everyday Examples

Long before computers were invented and prevalent in all aspects of everyday life, the principles of engineering geometry were used extensively in design development. The principles helped solve complex technical problems graphically rather than mathematically and create detailed technical drawings.

For example, parachutes were first invented in the late 1700s, long before airplanes and space flight (see figure 2.2). The ideal shape was calculated using principles of engineering geometry and has since been refined over time. Parachutes created today for use in aerospace use high power computing to optimize the parachute area needed to create the right amount of drag for the capsule or package to land at a safe speed (see figure 2.3).

Figure 2.2 - Early Parachute Design - Image courtesy of US Library of Congress

Figure 2.3 - Modern Parachute Design - Image courtesy of NASA

There are many examples of engineering geometry principles in architecture. While you can probably think of a number of structures that have domes, you may not have realized that many were designed with either a spherical or parabolic base. An example of a spherical done is the Pantheon in Rome, Italy (see figure 2.4), which is still the largest unreinforced concrete dome in the world.

Figure 2.4 - Pantheon Dome - Image courtesy of Sailko, Wikimedia

An example of a parabolic dome is the United States Capitol Building (see figure 2.5). The parabolic shape of the dome is especially evident when looking at the dome from the outside and especially when looking at a full-section view.

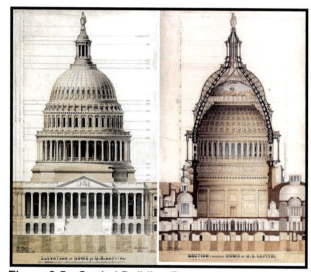
Figure 2.5 - Capitol Building Dome - Image courtesy of Architect of the Capitol

An example of a far more complicated dome that does not follow simple mathematics is the onion dome (see figure 2.6). Famous onion domes can be seen at St. Basil's Cathedral in Moscow, Russia. In some instances, this type of dome may be considered to have a freeform surface.

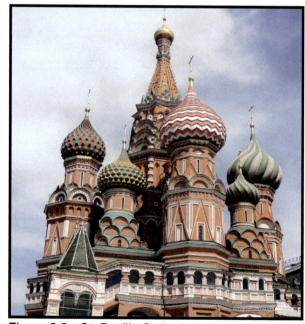
Figure 2.6 - St. Basil's Cathedral - Image courtesy of Zeynel Cebeci, Wikimedia

Engineering geometry has also played an important role in the technological advancement of scientific instruments. For example, a telescope relies on parabolic convex and concave surfaces to focus the magnified image. A refractor telescope is the simplest and was first used hundreds of years ago. The diagram in figure 2.7 shows the simplicity of the apparatus. As you can imagine, the first telescopes were quite primitive and made from simple materials.

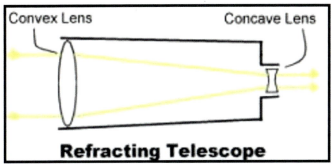

Figure 2.7 - Refracting Telescope - Image courtesy of NASA

Coordinate Systems

The idea of coordinate systems, different types of surfaces, and geometric relations may have been covered in previous math classes, but now we will look at these items from a technical graphics perspective. The main coordinate systems we will examine are Cartesian, cylindrical, and polar as each has applications in CAD modeling. Using the coordinate system helps locate a point in space. Engineering geometry allows for the use of various geometric relations between points, lines, and planes ,which are extremely useful for creating constraints need for efficient 3-D modeling.

This unit will examine the main different types of surfaces such as single- or double-curved, as this may help you determine how to go about constructing certain objects in CAD-based systems. This unit will examine the three main types of Boolean operations you may employ to create a wide variety of objects.

From math classes you may remember seeing a variety of coordinate systems with the main ones being Cartesian (most common), polar, cylindrical, and spherical. The Cartesian coordinate system is based on locating a point in space in all three dimensions and uses the familiar x-y-z coordinate system. The main difference between a Cartesian (rectangular) system and the other coordinate systems is the addition of using one or more angles rather than just distance to define a location (see figure 2.8)

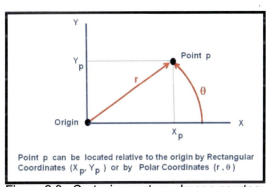

Figure 2.8 - Cartesian system - Image courtesy of NASA

Let's think of these coordinate systems based on how they could be used to build a robot. Each coordinate system lends itself to varying strengths in different types of movement or measurement. Figure 2.9 shows a large assembly line robot that is used to help join various materials in a NASA laboratory.

Figure 2.9 - Assembly Line Robot - Image courtesy of NASA

The following are examples of robots which use different coordinate systems.

Cartesian - Let's start with simple up-and-down or back-and-forth motion that goes along any of the main three perpendicular axes (x-, y-, or z-axis). In this case, we would build a robot based on the Cartesian coordinate system (see figure 2.10). This works well for actions such as picking a box off the shelf like automated machines in a warehouse or other simple tasks that do not require any circular motion.

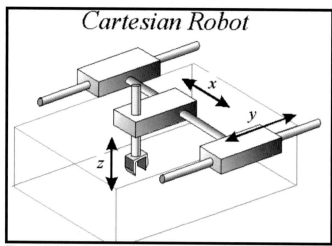

Figure 2.10 - Cartesian Robot - Image courtesy of NASA

Cylindrical - While a robot built on a Cartesian coordinate system may work for limited, simple situations, expanding to use a robot built on a cylindrical coordinate system introduces rotational motion, as points in space are now defined by two distances and an angle. This would allow us to add a swivel arm as seen in figure 2.11. By using distances and an angle you can now do more complex actions and you see many robots like this on an assembly line. Note how this type of robot operates within a contained box.

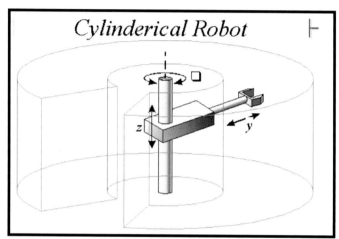

Figure 2.11 - Cylinderical Robot - Image courtesy of NASA

Polar - The next level coordinate system is a polar (or spherical) that is based on one direction and two angles. A robot built based on this type of coordinate system has the most flexibility and can be used for complex tasks such as welding irregular surfaces. The cylindrical outline shows the area in which the robot can operate. Though it may seem more limited than with the cylindrical coordinate systems, it is minor compared to the twisting and more complex movements that are possible with a robot built with the polar coordinate system.

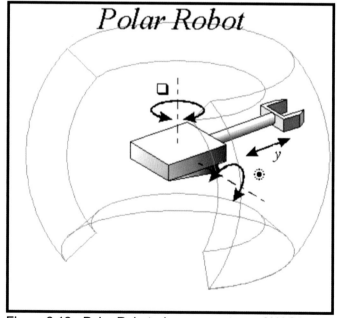

Figure 2.12 - Polar Robot - Image courtesy of NASA

Geometric Relations

Let's explore common geometric relations using simple games you probably have played at some point in your life already. Let's start with the game, Tic-Tac-Toe. Chances are you could quickly draw the cross-hatched playing board but you may not have realized how many useful geometric relationships could be identified as demonstrated in figure 2.13.

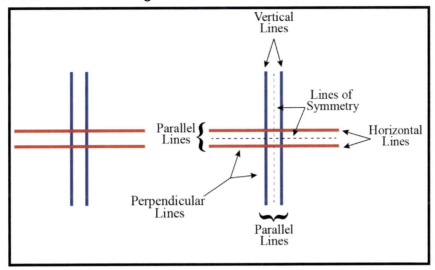

Figure 2.13 - Geometric Relations

A simple dartboard is another great example to show geometric relations (see figure 2.14). In this case, the two main geometric relationships are concentric circles and shape symmetry. Concentric means having the same center point (in this case the bullseye). On a dart board, the rings are concentric. You can clearly see two rings concentric to the bullseye marking the different point levels when a dart hits the board.

Figure 2.14 - Dart Board - Image courtesy of Andy Roberts at Wikipedia

Figure 2.15 - Refracting Telescope - Image courtesy of AlphaOrionis42 at Wikipedia

Engineering geometry principles also have played an important role in the technological advancement of scientific instruments. For example, a telescope relies on parabolic convex and concave surfaces to focus the magnified image. A refracting telescope (see figure 2.15) is the simplest of these types of apparatuses and was first used hundreds of years ago. As you can imagine, the first telescopes were quite primitive and made from simple materials.

Surface Types

While there are many different classifications used to describe surfaces in engineering geometry, we will examine the most common types: planar, single-curved, double-curved, and freeform.

A planar surface is essentially a closed geometric shape on a plane with no thickness. In CAD-based systems, you draw various planar shapes with the intention of extruding or revolving them to create a solid model. Since real world objects have some sort of thickness associated with them, think of a planar object as something like a stop sign (see figure 2.16). It has a closed geometric shape (octagon) and the thickness is thin compared to the surface area of the front.

A single-curved surface yields a cylinder or a cone. Think about starting with a planar circular base and then drawing lines upwards from that base. If the lines are parallel you get some form of cylinder (see figure 2.17), if they converge to a single point you create a cone. Single-curved surfaces are "developable," which means they can be translated onto a planar surface without distortion.

The simplest double-curved surface is a sphere, such as a basketball. A double-curve surface has no straight lines and the bounds of the object are curved. Another shape considered double-curved is an ellipsoid, or rather a squashed sphere like the pumpkin in figure 2.18.

Double-curved surfaces are "undevelopable," meaning they cannot be translated to a planar surface unless there is distortion. Think about trying to unroll a map of the world to a flat surface - it can not be done without distortion.

Figure 2.16 - Stop Sign

Figure 2.17 - Canned Food

Figure 2.18 - Pumpkin

A freeform surface is one that cannot be described with simple mathematics and does not follow any discernible pattern. Many marble statues you see at museums are great examples of a freeform surface. The inexpensive plaster cast prototype in figure 2.19 has reference nails placed over the face and neck region. This allows the artist to take a number of measurements and translate the design to more expensive marble. This takes advantage of knowledge of the Cartesian coordinate system to translate an elaborate freeform design, which shows how engineering geometry can be applied in unexpected ways.

Figure 2.19 - Plaster Cast Prototype

Gumball Machine

A basic gumball machine is pictured in figure 2.20. What are the primitive shapes needed to make a rough draft of this object in a CAD-based modeling program?

While there are various options, here is one possible solution: 6 solid blocks (rectangular prisms), 2 of which have additional features (holes), and 1 hollow glass sphere. Note that curves to the edges on the top and bottom pieces can be added in CAD after the basic shapes are made.

Figure 2.20 - Gumball Mahine

Extrude vs Revolve

Extrude - There are certain shapes that can be created by either an "extrude" or a "revolve" feature in CAD modeling. In the case of an extrude, a planar shape is extruded a certain distance to give a solid shape. Take for example, the classic pink eraser in figure 2.21. The outline on the left extruded gives the shape on the right.

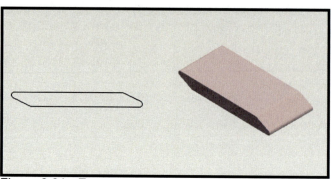

Figure 2.21 - Eraser

Revolve - In other instances, an outline can be made and then revolved around an axis to create a solid shape. This is the case with the candlestick in figure 2.22.

In some situations, either method can be used to create the object. See the figure 2.23 for making a simple tube. Depending on the design intent, you may need to use one method over the other when creating an object, even when both ways will give you the same result.

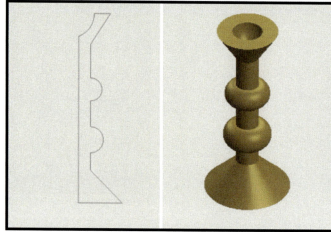

Figure 2.22 - Candlestick

Figure 2.23 -Tube (left is using extrude and right is using revolve)

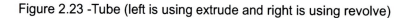

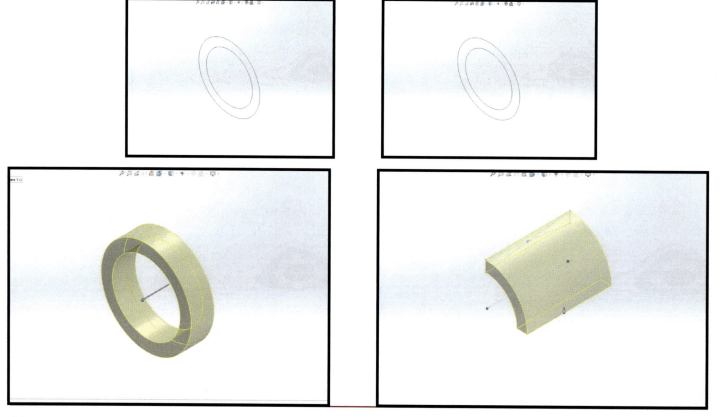

Foundational Engineering Graphics: Principles and Applications

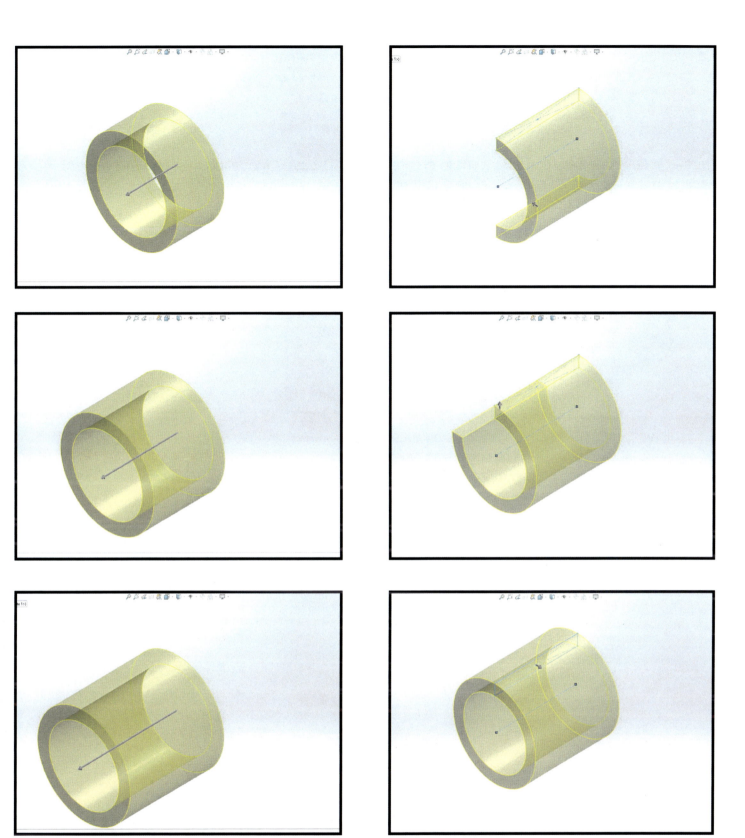

Reflection Question

Can you think of a time where you used a graphical solution rather than a mathematical solution to solve a complex math problem? You may have done these types of problems in geometry without realizing what methodology you were indeed using to derive an answer.

Quiz

Here is a set of objects that you want to create in a CAD-based modeling system. You want to make each object in as few steps as possible. In other words, just use one feature. Would you use an extrude, revolve, or are they equally valid methods?

A

B
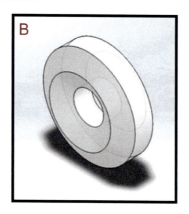

C
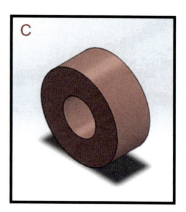

1. Which object can only be created using extrude?
2. Which object can only be created using revolve?
3. For which object could you use extrude OR revolve?

Solutions in Appendix

UNIT 3

ORTHOGRAPHIC PROJECTION

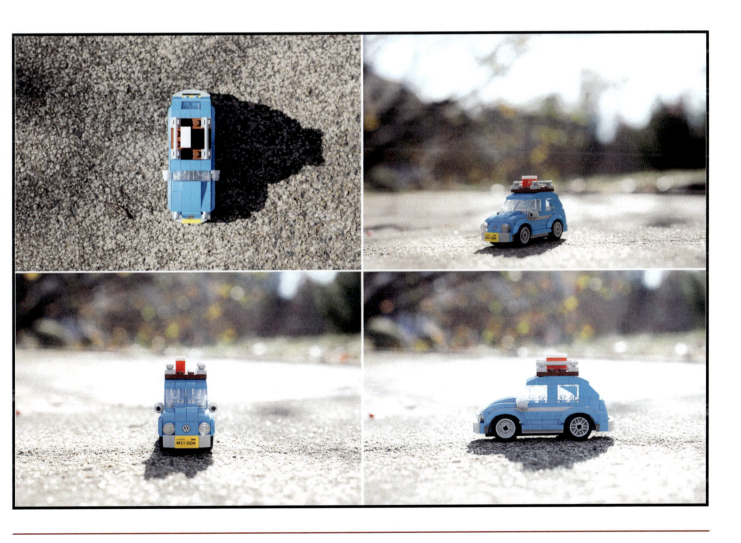

Introduction

Orthographic projection is a cornerstone concept in technical graphics that requires an understanding of projection planes and different types of surfaces. It gives us a way to represent a 3-D object on a 2-D sheet of paper. The topics in this unit are important for creating technical drawings to communicate designs to a wide audience. In orthographic projection, an object can be represented as an axonometric projection where you see all three sides of an object as a single view. Alternatively, it can be displayed as a collection of drawings arranged in a specific way, where you see only two dimensions per view (multiview) as in figure 3.1. In this unit, we will focus on the creation and interpretation of these multiview drawings.

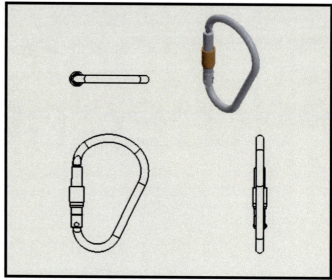

Figure 3.1 - Carabiner Multiview

Everyday Uses

You may be familiar with online buy-trade-sell websites where sellers post images of an object with the hopes of finding an interested buyer. This type of platform is especially popular for large items like automobiles, boats, or motorhomes. The images typically posted for a car often mimic the multiviews of orthographic projection. While one image usually shows the whole car at an angle, subsequent images may show just the side (profile), the front (grill and hood), and on occasion from the top–especially if the seller wants to highlight a feature like a sunroof or roof rack to potential buyers. While these images are often arranged in a linear format, generic photo grid, or as a slideshow for ease of viewing on mobile devices, we can take these images and arrange them akin to a multiview drawing based off of orthographic projection theory as in figure 3.2.

Figure 3.2 - Multiview Layout

Another example of multiview projection in use is at a local science museum where visitors are encouraged to examine delicate objects encased in a glass box. This allows clear viewing of the various sides in a way that is congruent with orthographic projection. Let's look at some examples from a local museum in figure 3.3. The bird's nest is shown from the front view, the rat skeleton from the top view, and the jaws of a shark ancestor appears in a side (profile) view.

Figure 3.3 - Exhibit VIews

Basics

Frequently, engineers and designers have ideas they need to show others while they are generating potential solutions to a problem. One way to demonstrate that a design solution may be feasible is by building prototypes in 3-D modeling software that allows you to go back and forth from an isometric view to multiviews. Multiview drawings are particularly important because they form the basis for communicating the size of an object. They are then annotated to show additional information, such as dimensions or manufacturing instructions. They are often used in patent applications like the top, front, and bottom image of the jar in figure 3.4. Due to well-known established standards for technical drawing, it is possible to read and interpret technical drawings from many, many years ago.

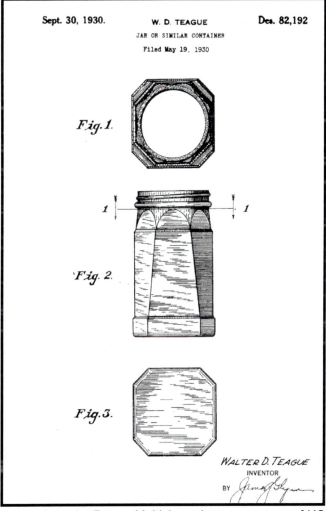

Figure 3.4 - Jar Patent Multiview - Image courtesy of US Patent and Trademark Office

Theory

Orthographic projection theory can be challenging for many students because it requires the ability to translate a 3-D object onto a 2-D sheet of paper. This unit will focus on the creation and interpretation of multiview projections that you can think of as a combination of many drawings arranged in a particular way. Previously in the sketching unit, we introduced regular grid paper (used for multiview drawings) so you may have already gained some exposure to these types of technical drawings even if you were unsure of the terminology.

In this unit, we will examine why multiview drawings are oriented in a certain way (see figure 3.5), the dimensions shown in each view, and techniques to create these drawings by hand. Multiview projection drawings are a critical component of many other technical graphics topics including working drawings, section view drawings, auxiliary drawings, and dimensioning. This unit will also explore the three main planes of projection, glass box theory and six principle views, surface identification, and how to use multiview projection drawings to construct 3-D images.

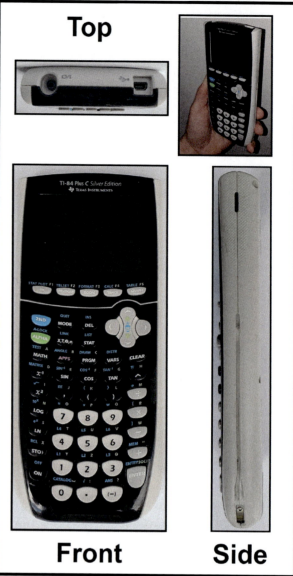

Figure 3.5 - Calculator Multiview

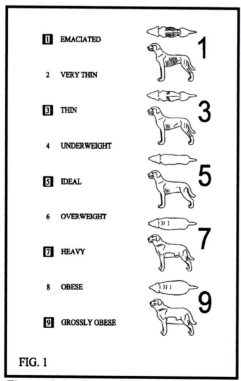

Figure 3.6 - Dog Shape Multiview - Image courtesy of US Patent and Trademark Office

Another practical example of using multiview projection drawings actually occurs in a vet's office. To help assess the health of a dog, a vet will not only examine the front view, but also the top and side views to make clinical assessments. Often, a drawing with the different views shown in figure 3.6 is present in the examination room as a reference for dog owners to examine and to help veterinarians better explain the shape of the animal they are examining.

When one observes their surroundings they see objects from multiple perspectives. Perceiving objects from different planes of view (top, front, and side) provide different details. Being able to observer details from different views allows for clear communication when describing the object. For instance, viewing landmarks, such as the Statue of Liberty (see figure 3.7) from multiple perspectives provide various details. When looking from the front and side, one can view great details of the statue. However, viewing the statue from the top provides details of the shape of the base of the statue.

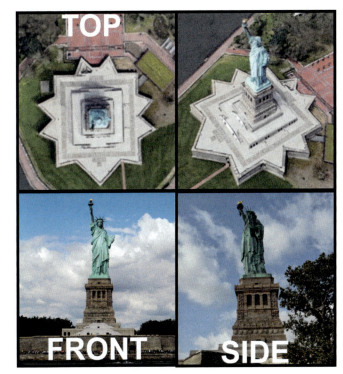

Figure 3.7 - Statue of Liberty Multiview - Image courtesy of Google Maps (top two images) and Andrew Maiman at Wikimedia (bottom two images)

Multiview projection stems from parallel projection (see figure 3.8). In parallel projection, the lines of sight from the view to the object are parallel (do not converge), no matter the distance between the observer and the object, the projection of the object will always be projected at the same size onto the projection plane in true size and shape. In perspective projection, the object will appear as we see it in real life, meaning that the object will appear to change size depending on the distance from the observer

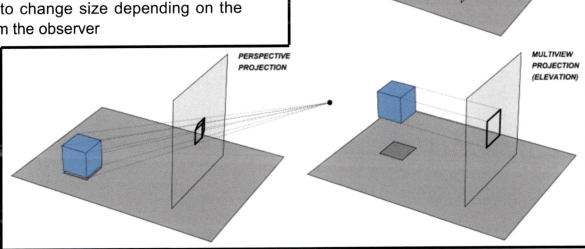

Figure 3.8 - Parallel vs Perspective Projection - Image courtesy of SharkD at Wikimedia

Planes of Projection

We examined the order of the observer, the plane of projection, and the object in the theory section above. Terminology is important here, as many CAD systems also use this terminology to help you orient your 3-D drawing. Now, let's look at the cases of frontal, horizontal, and profile planes of projection relative to an everyday object: a chair. We are going to look at the relationship between the front view and the frontal plane, top view and the horizontal plane, and right side view and profile plane.

Figure 3.9 has a photograph of the front view of a child's chair. We are going to project the visible lines from the chair (the object) to the frontal plane. We will then make a drawing of what we see on this frontal plane, which will be the front view of a multiview projection drawing. In this view, we see the width and height of the object.

 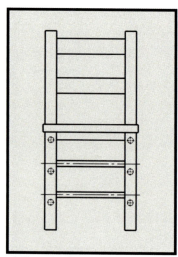

Figure 3.9 - Chair Front View (Photo and Drawing)

Do you see the connection between the front view of the chair and the front view of a technical drawing? What does the technical drawing include that you do not see in the photograph?

Next, let's look at the top view of the object as a CAD drawing. We are going to project this view to the horizontal plane. We will now make a drawing of what we see on this horizontal plane, which corresponds to the top view of a multiview projection drawing. In this view, we see the width and depth of the object.

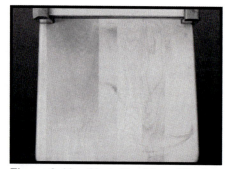 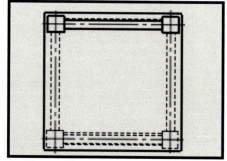

Figure 3.10 - Chair Top View (Photo and Drawing)

The technical drawing of the chair includes annotation showing the legs and rungs below the seat of the chair. These features are hidden in the photograph of the top of the chair. This highlights how more information can be given in orthographic multiview drawings rather than just images of an object taken with a camera.

We will follow the same steps while looking at the profile plane and constructing the right side view. Could you have made this drawing on your own after seeing the connection between the photograph and the technical drawing for the front and top views? What dimensions do we see in this view? We see the height and depth of the object here.

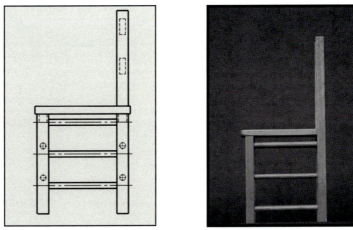

Figure 3.11 - Chair Side View (Photo and Drawing)

Precedence of Lines

Every multiview projection drawing needs to show visible edges on the normal surface and correct annotation for hidden features, as well as circles/arcs. The three major line types in order of precedence are: visible, hidden, center line/mark. Each line type is drawn using industry-wide accepted notation. Visible lines are thick and dark. Hidden lines are lighter and dashed with equal spacing. Center lines are lighter and have an alternating long-short-long line type. Center marks have a hash mark and then radiating lines to the edge of the arc or circle. The best way to review this is to look at some examples such as in figure 3.12.

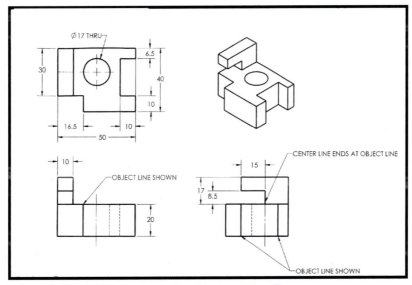

Figure 3.12 - Multiview Including Various Line Types

Glass Box Theory

Orthographic projection and creating multiview projection drawings can be a challenging topic for students, but many find it helpful to learn about Glass Box Theory. We will use this as our base to build on throughout the rest of the content unit. You may have seen collectible objects displayed like the cell phone in the box in figure 3.13 to keep them in pristine condition.

Figure 3.13 - Cell Phone in Glass Box

According to Glass Box Theory, the surfaces of the glass box represent the different normal planes of projection:

- Frontal (corresponds to front and back views)
- Horizontal (corresponds to top and bottom views)
- Profile (corresponds to right side and left side views)

You often see the planes listed like this in CAD-based software. You are able to draw planar profiles directly on any of the surfaces to start your 3-D model.

To get any of the six principle multiview projections, observe what you see at that face of the box. For instance, to determine the front multiview projection of the phone, position yourself parallel to the "frontal plane" according to Glass Box Theory (see figure 3.14). Now what you see is the front view of the phone, and you can translate that into a technical drawing. A similar process is done to derive the other five principle multiview projection drawings.

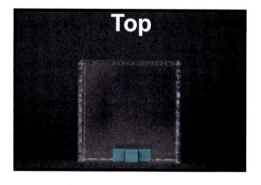

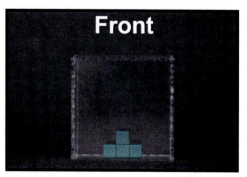
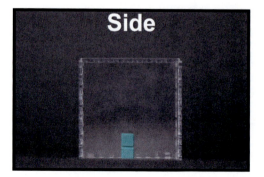

Figure 3.14 - Multiview in Glass Box

Third Angle vs First Angle Projection

You may be wondering how to orient and align all these technical multiview projection drawings relative to each other. In Glass Box Theory, the way we arrange multiview drawings directly corresponds to how we unfold the box while looking directly at the front view.

In the United States the Glass Box is unfolded so the multiviews are arranged as shown in the multiview projection image in figure 3.15. This arrangement of the six principle views is called Third-Angle Projection. This type of arrangement is also denoted with a symbol as shown in figure 3.15. There is also First-Angle Projection as shown in figure 3.16 that is widely used in India and European countries.

Third Angle Projection

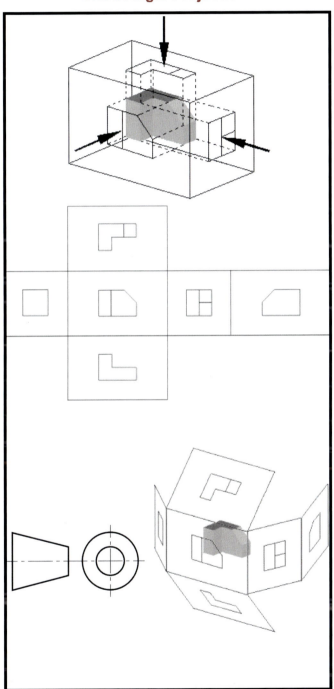

Figure 3.15 - Third Angle Perspective - Image courtesy of Maksim - Wikimedia

First Angle Projection

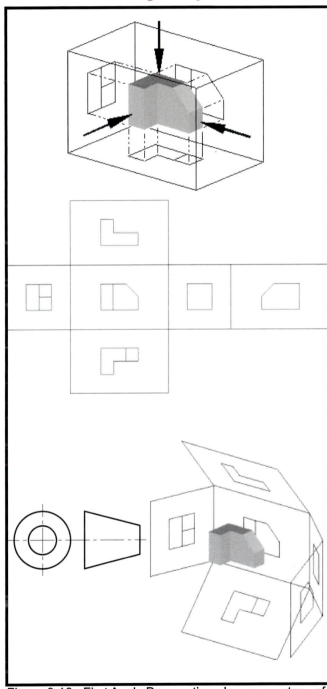

Figure 3.16 - First Angle Perspective - Image courtesy of Baxelrod - Wikimedia

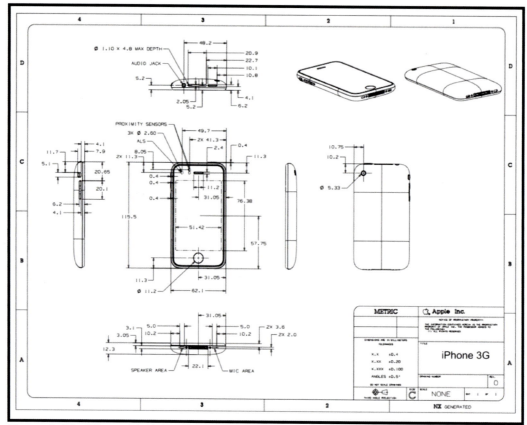

Figure 3.17 - iPhone Multiview Technical Drawing - Image courtesy of SolidSmack

We've looked at images of the phone and how it looks from different views. Now let's look at a corresponding technical document. The image in figure 3.17 was created by Apple for third-party vendors so they could develop and create various cases and accessories for an early generation iPhone.

Can you identify the front view? Top view? Right side view?

While we are focusing on only the views themselves, you also see how multiview projection drawings are used in other important technical drawing topics such as dimensioning, section views, and working drawings. Mastering this unit will help you in other topics so if needed, come back and review this unit as often as necessary.

Reflection Questions
1. Have you ever researched an item online where you examined it from different views? What was the item? How did seeing the different views influence your decision to buy or not buy the item?

2. What objects have you examined in a "glass box" (or plastic box)? This may have been at a museum, a store, or your home. Think about items you may have been told to "look at but don't touch" if you are having a tough time coming up with an example.

Activity

1. Find an object nearby that can be held in your left hand. After examining it, identify what you think would be the best front view (most descriptive). While looking at the front view, rotate the object towards you to examine the top view. Now, rotate the object back to the front view. To look at the right side view, turn the right side of the object towards you.

2. Let's make a multiview drawing based on the object in the glass box in figure 3.18. Get a sheet of rectangular grid paper. We will go through the process to make a multiview drawing together. Let's assume that the height = width = depth for the object.

Figure 3.18 - Object in Glass Box

Begin by determining the overall dimensions.
Next, sketch in the visible lines and main features.
Finally, add items like hidden lines and center lines or marks.

Step 1:

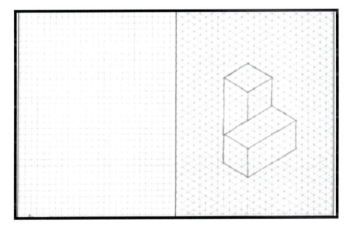

Step 2:

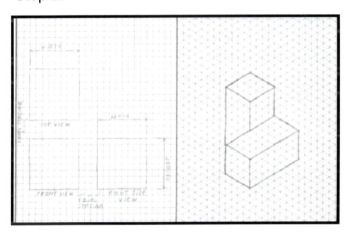

Step 3:

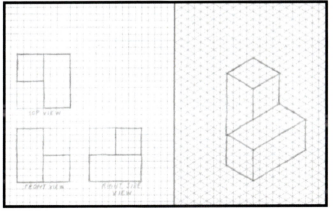

Step 4:

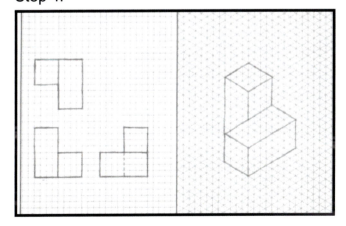

Quiz 1

We've talked about planes of projection and reviewed precedence of lines. Given these objects, select the items that are correctly annotated.

1. Does the drawing in figure 3.18 have the correct marking and lines? Hint: First check visible line, then hidden lines, and then centerline or marks or for each view.

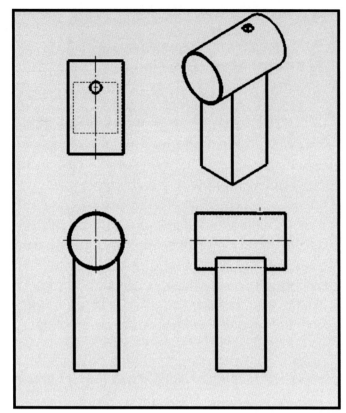

Figure 3.18 - Quiz - Object One

2. Does the drawing in figure 3.19 have the correct marking and lines? Hint: First check visible line, then hidden lines, and then centerline or marks or for each view.

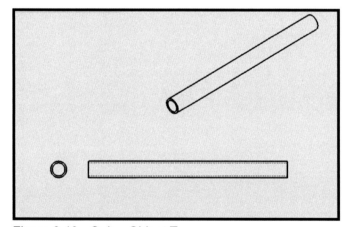

Figure 3.19 - Quiz - Object Two

3. Does the drawing in figure 3.20 have the correct marking and lines? Hint: First check visible line, then hidden lines, and then centerline or marks or for each view..

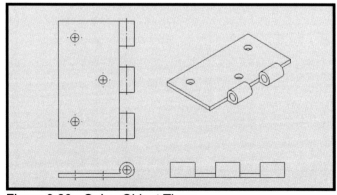

Figure 3.20 - Quiz - Object Three

Quiz 2

Before jumping into theory, let's first practice identifying the correct top, front, and right side view of common objects. You should select the most descriptive view as the front view of the object. You may notice that some objects only need two views because of the presence of a line of symmetry in the object.

For each row of images, identify which is the most descriptive view of the object and should be used as the front view.

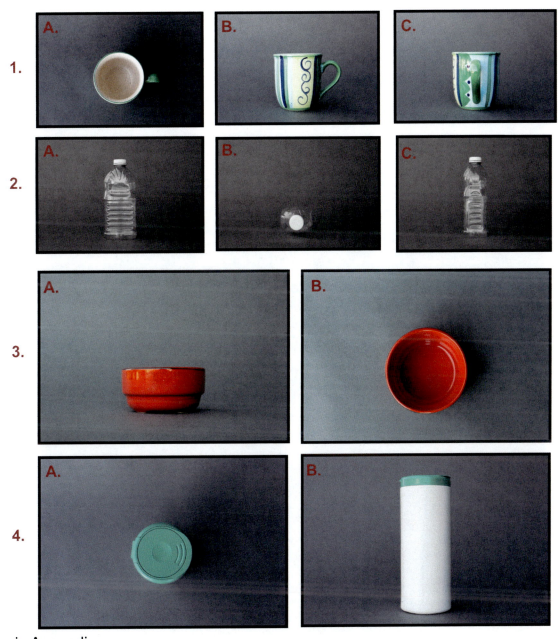

Solutions in Appendix

UNIT 4

PICTORIAL PROJECTION

Images courtesy of NASA

Foundational Engineering Graphics: Principles and Applications

Introduction

Of all of the topics in technical graphics, pictorial projection is the area you most likely have had the most exposure to in everyday life. Pictorial projections are the 2-D representations of 3-D objects that you see on paper or a computer screen. The image of the nail clippers in figure 4.1 below demonstrates this type of layout where you seem to see the front, top, and right side of the object in a single image. Pictorial projections are used extensively in advertisements, presentations, and brochures–practically any visual material looking to communicate a design. Pictorial projections are typically used by engineers and designers to help explain their ideas to project collaborators, demonstrate prototype concepts, and communicate with non-technical audiences. While there are different types of pictorial projections, the main categories include isometric (a subset of axonometric from orthographic projections), oblique, and perspective.

Figure 4.1 - Nail Clippers

Everyday Uses

There are many uses for pictorial projections. They are used to show the general public technical concepts in an easy to understand visual manner while often allowing engineers and designers to bypass what may be confusing technical jargon. These types of images usually are also incorporated into patent applications to show the new idea. Figures 4.2 and 4.3 are some sample pictorial projection images from various patents for a few items you probably have encountered at some point in your life.

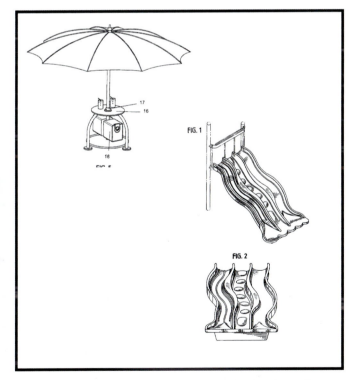

Figure 4.2 - Patent Sample Pictorial Projection - Image courtesy of US Patent and Trademark Office

Figure 4.3 - Carousel Mechanism - Image courtesy of US Patent and Trademark Office

Pictorial Projections

Pictorial projections are also used to show design concepts during the engineering design cycle. This is when a product is still in the modeling stage and has yet to be made into a prototype or tangible object. For example, there is recent interest in developing new models of aircraft. Figure 4.4 shows some concepts of how airplanes may look in the future. Do you see any advantages to these prototype models? What about disadvantages?

Figure 4.4 - Concept Aircraft - Images courtesy of NASA

Bookshelf Assembly Instructions

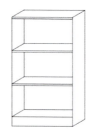

This image is an oblique perspective. An oblique perspective is another pictorial perspective.

Figure 4.5 - Bookshelf Oblique

Pictorial projections are common in the engineering design process and in everyday representations to visually communicate design ideas and products to a wide audience (see figures 4.4 and 4.5). Frequently, you create the pictorial projection of an object in CAD-based software using principles from engineering geometry. While pictorial projections may be a better visual communication tool, they are not prefered for dimensioning and annotation for manufacturing. That is why you often see pictorial and multiview projection drawings used together.

Figure 4.6 Dinosaur Skeleton

Isometric Projection

Isometric projections are a subset of axonometric projections stemming from parallel perspective theory. This means that no matter how far you are from an object, it will always appear to be projected in true size (see figure 4.6). The root "iso" means "equal." Going back to the idea of the Cartesian coordinate system in engineering geometry, we see the x-, y-, and z-axes at a viewpoint where they appear at 120 degree angles to each other. This angle is easy to see if you look at isometric grid paper as seen in the Sketching Unit. The spacing of the grid is equal.

There are a few major annotation differences to be aware of when working with isometric pictorial projections. Hidden lines are typically not shown and centermarks are usually omitted in the isometric projection. While the viewer may get a better overall feel for an object, they do not always have all of the information necessary when it comes to manufacturing an object from the isometric pictorial projection alone.

This is the most common pictorial projection type of drawing used in technical graphics. While the object may appear skewed in a drawing because of parallel projection, the equal spacing allows for quick translation to a multiview drawing (see figure 4.7).

Many CAD models are created as an isometric projections making this by far the most dominant pictorial projection technique. Did you know that isometric pictorial projections can be formatted to be sent directly to a 3-D printer? The applications for 3-D printed objects is rapidly expanding and you may have had some experience with this technique already. For example, an isometric pictorial projection of a ratchet was created in a CAD program and 3-D printed on the International Space Station as seen in figure 4.8. Astronauts can now make tools on demand giving more flexibility in planning and abilities to solve more unforeseen challenges.

Orthographic Projection to Isometric Projection Drawings

In the Orthographic Projection Unit we focused on making multiview projection drawings from an object or isometric drawing. In this unit, we will do the opposite. Given a set of multiview projection drawings, you can create the isometric pictorial. It is important to learn how to go back and forth between these two projection systems. This may seem awkward at first. With practice, you will be able to go back and forth between the two types of parallel projection drawings with relative ease.

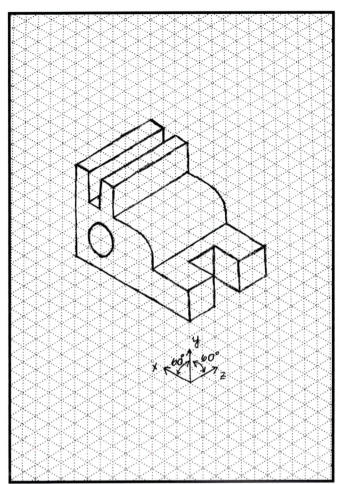

Figure 4.7 - Isometric Pictorial Projection

Figure 4.8 - Isometric Projection of 3D Printed Ratchet - Image courtesy of NASA

The type of paper used to create these isometric drawings is isometric grid paper. There is a sample in the Appendix. You had practice drawing on this paper in the Sketching Unit but may not have been introduced to a technical name for this type of drawing.

As a refresher, recall the arrangement of a multiview drawing with the object oriented as an isometric pictorial as in figure 4.9. The idea of this section is to take that top, front, and right side view and construct the isometric pictorial. The tiger is a difficult object to draw by hand, so we will use work with a basic object to demonstrate this process.

Let's start by outlining top, front, and right side views.

In each of the multiview drawings, you see two of the three major dimensions (height, width, and/or depth). See figure 4.10.

Step 1 - The first general step in creating an isometric drawing is to determine the correct bounds of the object and draw a construction box on the isometric grid paper. See figure 4.11.

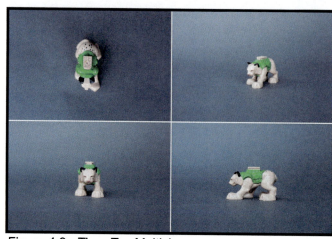
Figure 4.9 - Tiger Toy Multiview

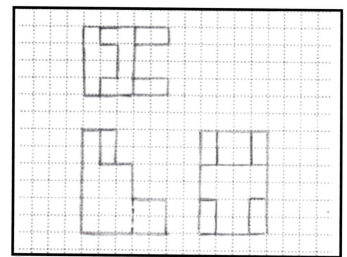
Figure 4.10 - Multiview Drawing

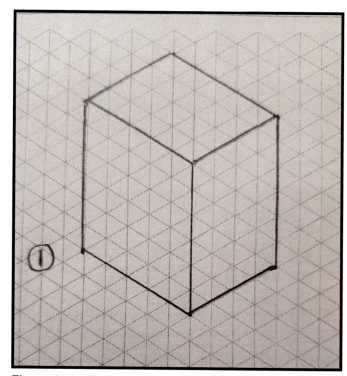
Figure 4.11 - Object Bounds

Step 2 - The second step is to identify normal surfaces, starting with those at the bounds of the object. This is the hardest step of the process as you need to go back and forth between at least two of the multiview drawings to place visible lines correctly. This step takes practice, so don't get discouraged if it takes some trial and error to get the object correct. Some students find it easier to relate each vertex of the object to a Cartesian coordinate system. While we will not go into detail of that here, you may try this method on your own. See figure 4.12.

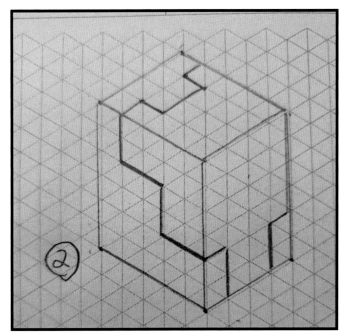

Figure 4.12 - Identify Normal Surfaces

Step 3 - The third step is to identify internal features such as cut-outs or holes and incorporate those into your isometric pictorial. This step may or may not be needed depending on your object. See figure 4.13

Figure 4.13 - Internal Surfaces

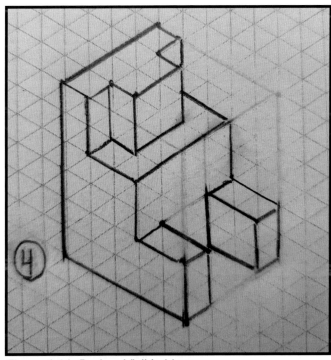

Figure 4.14 -Darken Visible Lines

Step 4 - The last step is to darken the visible lines so you clearly show the isometric pictorial from the orthographic multiview drawing. See figure 4.12.

You may find it helpful to label the surfaces in the multiview drawings and make sure they correspond correctly in the isometric projection. Figure 4.14 is the same example from the previous page with addition of color coding. The white surface in the finished pictorial projection is the hidden surface indicated in the front view of the object.

In everyday practice, you will typically go from an isometric pictorial to the multiview drawing. Doing exercises going the other way, like you have done here, will help you develop stronger spatial skills and make it easier for you to interpret more intricate technical drawings in the future.

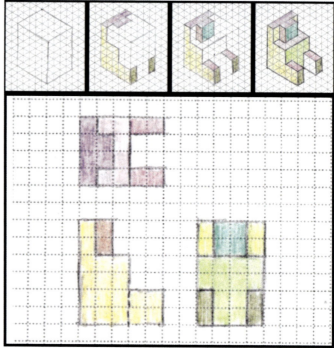

Figure 4.14 - Identify Normal Surfaces

Reflection Questions

Think of a time you have seen a pictorial projection. What was it? Have you ever used this type of drawing before for a project?

A common place that you might see pictorial projections is in assembly instructions. These graphics help users understand what the product will look like and how to fit the pieces together.

Can you think of any other uses for pictorial projections? The ideas mentioned so far include patents, product comparisons, and prototypes.

Activity

1. Construct a multiview drawing of the object in figure 4.15.

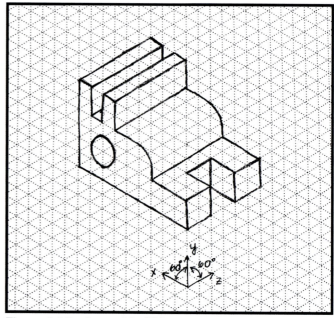

Figure 4.15 - Isometric Object

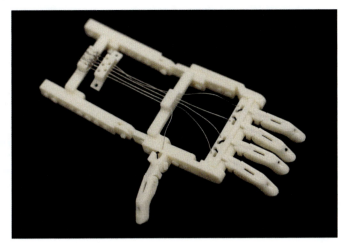

Figure 4.16 - 3-D Printed Prosthetic Hand - Image courtesy of U.S. Food and Drug Administration

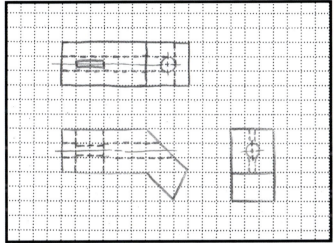

Figure 4.17 - Finger Aparatus Multiview (Top, Front, and Right Side)

2. One of the biggest advantages of 3-D printing is the ability to readily customize objects. While there are far-ranging applications, this can be especially important in medical applications such as prosthetics where a design will need to be adjusted to fit the patient. Traditionally, prosthetics are very expensive. However, 3-D printing can make a comparable model at a fraction of the cost. Hand prosthetics for children are widely available through a number of commercial companies. The image in figure 4.16 an FDA-approved hand prosthetic.

3. This design of a finger apparatus was originally created for children, but now we want to modify it for an adult user. In figure 4.17is the approximate plan for part of the adult finger apparatus (top, front, and right side) on grid paper. Your job is to create an isometric drawing of this elongated finger part. Note, for drawing purposes, the tip of the finger has been approximated to an inclined rather than rounded surface.

UNIT 5

WORKING DRAWINGS

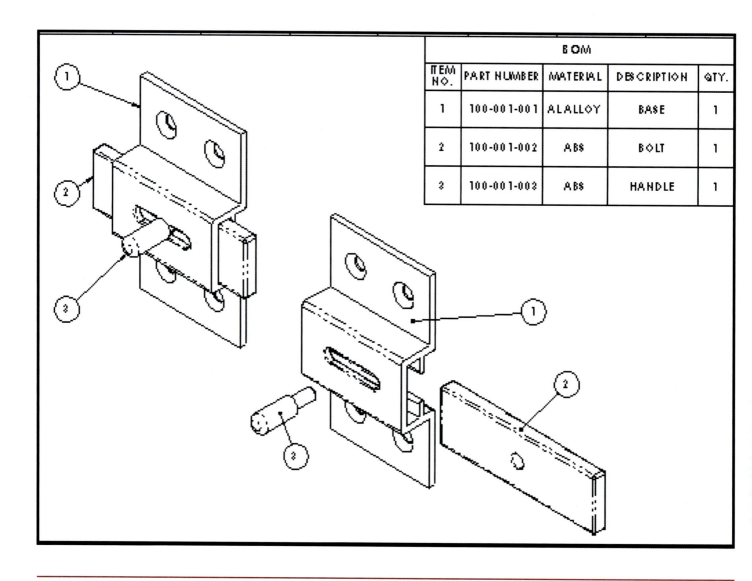

BOM				
ITEM NO.	PART NUMBER	MATERIAL	DESCRIPTION	QTY.
1	100-001-001	AL ALLOY	BASE	1
2	100-001-002	ABS	BOLT	1
3	100-001-003	ABS	HANDLE	1

Foundational Engineering Graphics: Principles and Applications

Unit 5 - Working Drawings

Introduction

The purpose of a working drawing set is to clearly communicate the manufacturing details and assembly instructions for an object as a whole assembly and for every unique part individually (see figure 5.1). Working drawing sets range from quite basic, like when making a prototype of a new design, to very large ones, such as those needed for patent applications. Regardless of complexity, working drawings all follow the same general pattern with specific standardized items. They are key for communication and play an active role in the revision process. Working drawings provide a methodology to document design changes in an organized manner. This section will discuss the various components needed for a complete working drawing set, accepted nomenclature, and examples for examination.

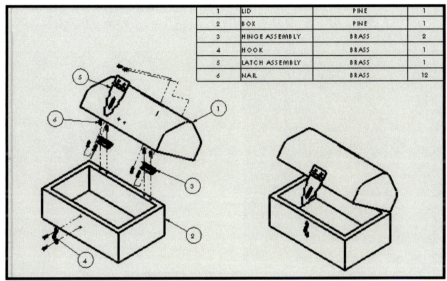

1	LID	PINE	1
2	BOX	PINE	1
3	HINGE ASSEMBLY	BRASS	2
4	HOOK	BRASS	1
5	LATCH ASSEMBLY	BRASS	1
6	NAIL	BRASS	12

Figure 5.1 - Jewelry Box Working Plans

Everyday Uses

You may have seen working drawings in your everyday life, but may not have known that a technical vocabulary existed to describe this set of information.

Have you ever put together a piece of furniture? Perhaps a bookshelf or a desk? Figure 5.3 depicts an example of a basic chair and the exploded view showing all the main parts needed for assembly.

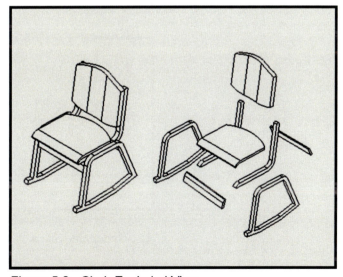

Figure 5.2 - Chair Exploded View

Chances are that piece of furniture came with an instruction booklet that showed you:
- Images of every unique part with a label (detail drawing)
- A list of all the materials, including fasteners, and number of each (bill of materials)
- Instructions with pictorial projections to communicate how the object ought to look at each step of construction
- An image of the entire object correctly put together (assembly drawing)

While all of the technical details of making each part may not have been included with this instruction booklet, the design is still clearly communicated. The assembler can then put together the furniture and use it as intended.

Working drawings are integral to documenting technological advancements in industry and construction. For example, patent applications are used to establish legal rights for an idea or technological advancement and rely heavily upon the inclusion of working drawings along with the necessary legal jargon. You may be surprised at the wide variety of objects and designs that are already patented.

Let's take a look at a recent invention that combines two common childhood favorites, a swing and a sprinkler, in an interesting way (see figures 5.3 and 5.4). You may have even tried something similar in the past but not as complicated.

Figure 5.3 - Swing   Figure 5.4 Sprinkler

Have you ever heard of a Waterfall Swing (see figure 5.6)? US Patent #8641544 details that very apparatus as seen in the example working drawings that show the entire unit and swing component below.

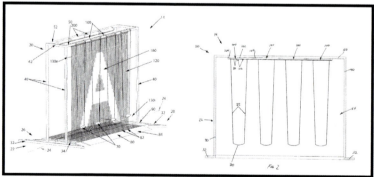

Figure 5.6 - Waterfall Swing Patent - Images courtesy of US Patent and Trademark Office

Space exploration remains at the forefront of technological advances. This industry relies heavily on the use of working drawings to document these feats of engineering. Many people are familiar with the American Space Shuttle Program that spanned multiple decades and can recall seeing videos of these enormous vehicles launching into space as in figure 5.7.

Every single component of that space shuttle is documented down to the exact number of nuts and bolts required for construction. It is this high level of detail in this working drawing set that allowed the initial construction of four of the exact same shuttles for the NASA Space Program. While the complete working drawing set is far too large to post in this unit, we can check out the NASA-created assembly drawing and exploded assembly drawing of the Discovery Space Shuttle in figures 5.8 and 5.9.

Figure 5.7 - Space Shuttle Launch - Image courtesy of NASA

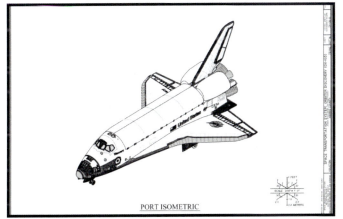

Figure 5.8 - Shuttle Assembly - Image courtesy of NASA

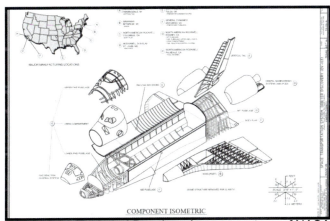

Figure 5.9 - Exploded Drawing - Image courtesy of NASA

As we saw above, working drawings are used in a variety of technically-oriented fields and are important for communicating designs to a wide variety of people associated with a project, from engineers to end users. Let's take time to dissect all of the components of a complete working drawing set that you will need to create or interpret while working on a design project.

Basics

Working drawings are used extensively in many fields of engineering and you can expect to see them used in industry to communicate designs, document design changes, and disclose design production information. To help you visualize a complete set of working drawings, let's examine what this would look like for a simple jewelry box (see figure 5.10).

Figure 5.10 - Jewelry Box and Working Drawing

Box Lid:

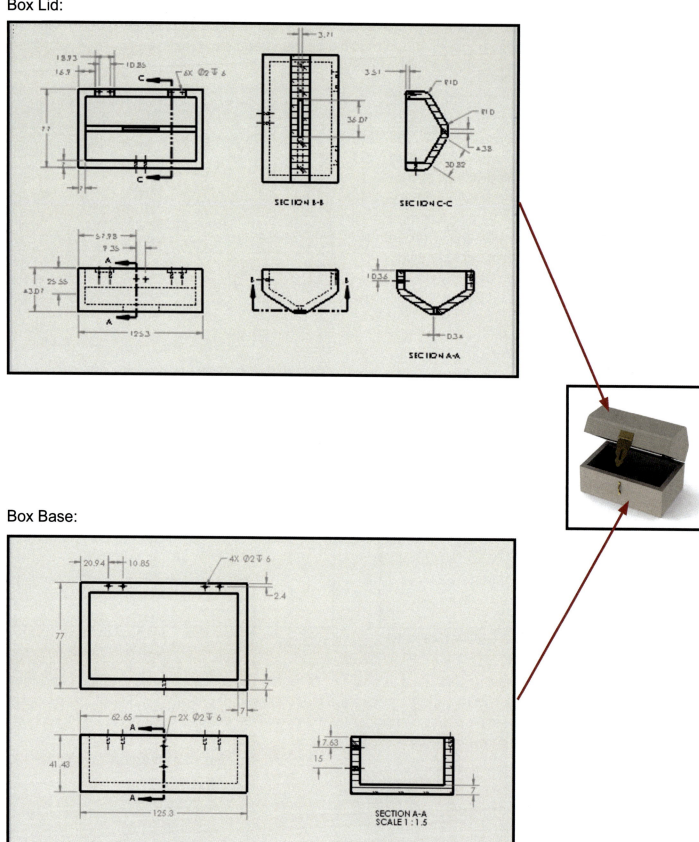

Box Base:

Lid Hinge:

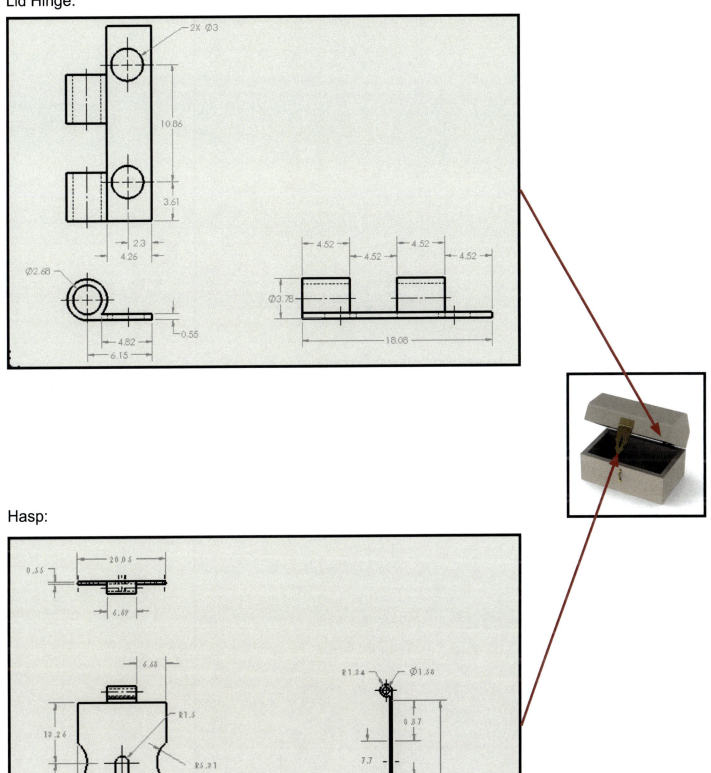

Hasp:

Hasp Hinge:

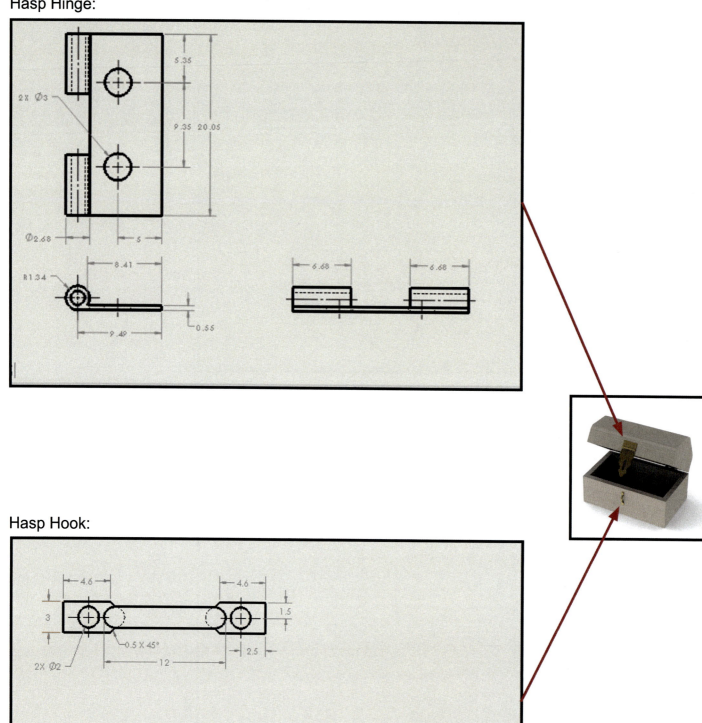

Hasp Hook:

Latch Rod:

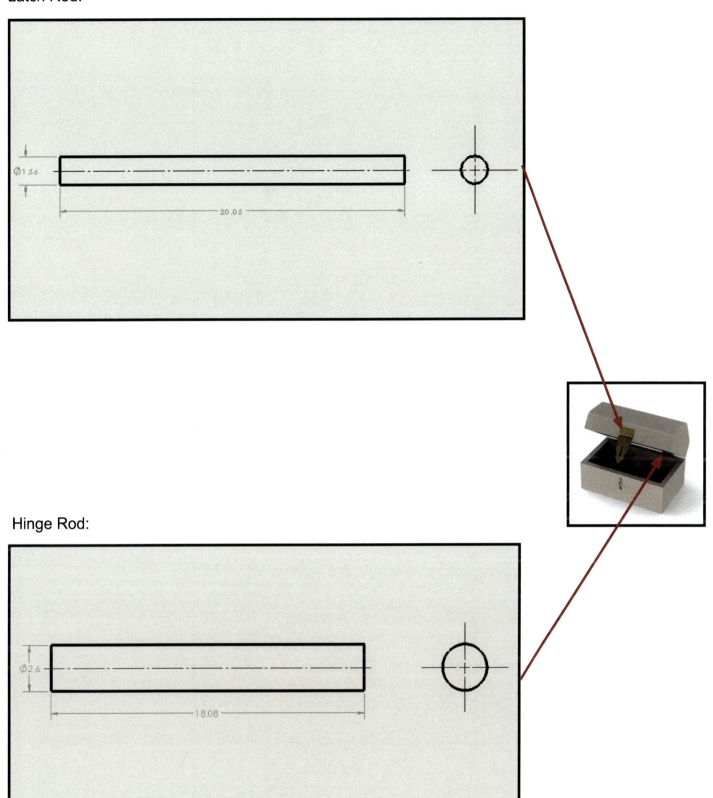

Ø1.56

20.05

Hinge Rod:

Ø2.6

18.08

Hinge Assembly:

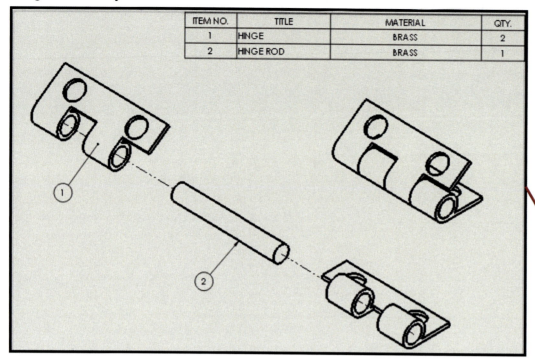

ITEM NO.	TITLE	MATERIAL	QTY.
1	HINGE	BRASS	2
2	HINGE ROD	BRASS	1

Hasp Assembly:

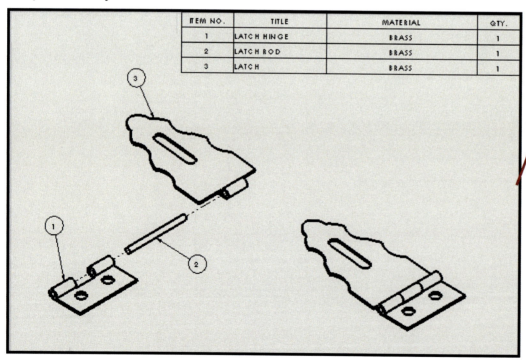

ITEM NO.	TITLE	MATERIAL	QTY.
1	LATCH HINGE	BRASS	1
2	LATCH ROD	BRASS	1
3	LATCH	BRASS	1

Box Assembly:

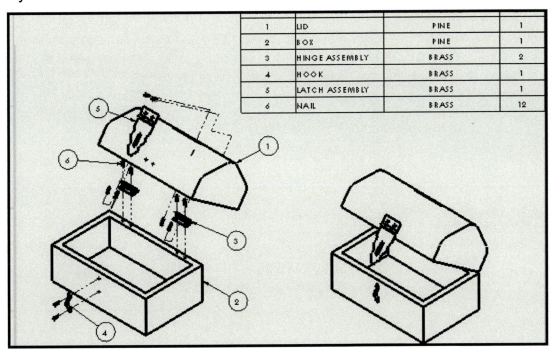

1	LID	PINE	1
2	BOX	PINE	1
3	HINGE ASSEMBLY	BRASS	2
4	HOOK	BRASS	1
5	LATCH ASSEMBLY	BRASS	1
6	NAIL	BRASS	12

Detail Drawing

Now we will examine a detail drawing (see figure 5.11) that is necessary for each non-standard (unique) part of the assembled object. Each detail drawing document shows information for just one part as an orthographic multiview projection. This allows for documentation of the shape, size (dimensions), material type, and any surface finish. All of which is required information for reproduction.

Detail drawings are often used by machinists or automated systems to create custom or unique parts for a design. Therefore, it is important that enough information is provided so the item can be reproduced many times in exactly the same way. This reproducibility is important to maintaining high levels of quality control and minimizing material loss. As an engineer, it is vital to create clear detail drawings that appropriately communicate the design to others, especially since the personnel working on the project or in the manufacturing process may change over time.

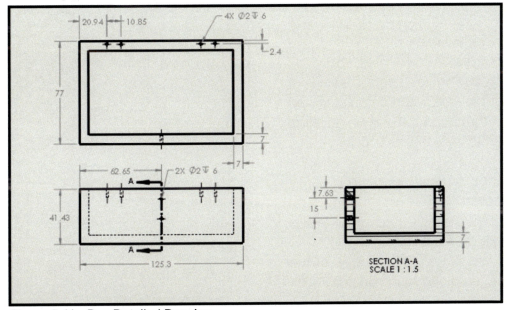

Figure 5.11 - Box Detailed Drawing

Title Block

Did you notice that all of the drawings in the jewelry box example had the same standard title block? A title block not only gives a cohesive, aesthetically-pleasing look to the set but also documents a variety of information. It is important to help keep track of documenting progress and revisions. While the title block template varies depending on the organization, it is typically located at the lower right corner of the drawing and contains the following information:.

- An identifiable title for the object
- Name of the company or organization
- Name of the creator(s)
- Date(s) of creation and revision
- Material(s)
- Scale and tolerances
- Angle of projection (often symbolic)
- Drawing number - often serves as an identifier when filing

Assembly Drawing

Next, let's examine an assembly drawing as part of the working drawing set. The main purpose of an assembly drawing is to show the operating position of the parts. In other words, how the parts should be set relative to one another. Assembly drawings also include:

- Bill of materials
- Item identifiers such as balloons
- Any specific machining or assembly instructions

Assemblies can be very simple, like the harmonica in figure 5.12, or very complex, like a NASA spaceship with thousands of parts. If you were looking to give a brief overview about your object to someone, most likely you would show the assembly drawing as it gives users a sense of how the item ought to look when completed and functional. This information can be understood regardless your audience's technical background.

Assembly drawings typically contain a bill of materials that shows the desired material and correct number of a specific part used in the design of the product. These are generally in a table format and are placed at an edge of the assembly drawing, as pictured in figure 5.13.

Figure 5.12 - Harmonica

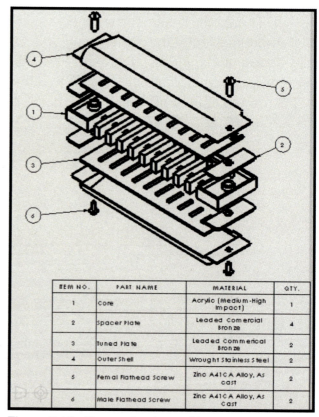

ITEM NO.	PART NAME	MATERIAL	QTY.
1	Core	Acrylic (Medium-High Impact)	1
2	Spacer Plate	Leaded Comercial Bronze	4
3	Tuned Plate	Leaded Commerical Bronze	2
4	Outer Shell	Wrought Stainless Steel	2
5	Femal Flathead Screw	Zinc A41CA Alloy, As cast	2
6	Male Flathead Screw	Zinc A41CA Alloy, As cast	2

Figure 5.13 - Harmonica Assembly

Standard Parts

Sometimes there are more parts in an assembly than the corresponding number of detail drawings. This is because standard parts such as nuts, bolts, screws, washers, gaskets, and various fasteners do not require detail drawings. Think of a standard part that you can pick up at your local hardware store or order from an online retailer because it is readily available, as displayed in figure 5.14. Information for these standard parts are often placed in the bill of materials for reference. Figure 5.15 shows a handful of sample standard parts you may recognize.

Figure 5.14 - Part Section of Hardware Store

Figure 5.15 - Standard Parts

Sample Set

Working drawing sets can quickly show the complexity of the object and all of the details needed to manufacture a product. Figure 5.16 shows a working drawing set for a slide bolt lock that you may find at a hardware store. Notice the section views included on the drawings for the individual parts in figure 5.16 on the following pages. This highlights the need for always using good dimensioning practices.

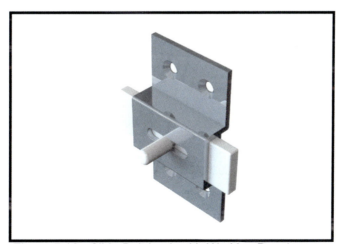

Figure 5.16 - Slide Bolt Lock with Working Drawings

Base:

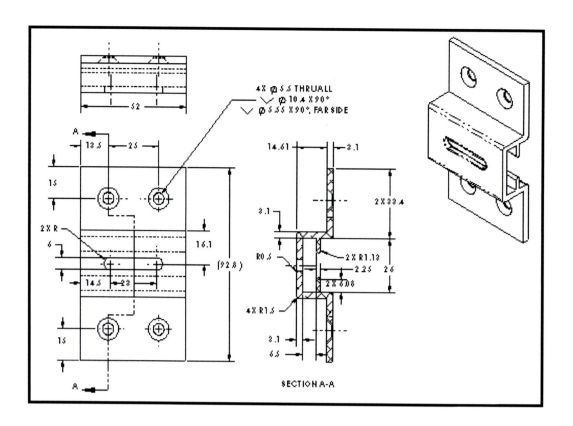

SECTION A-A

Bolt:

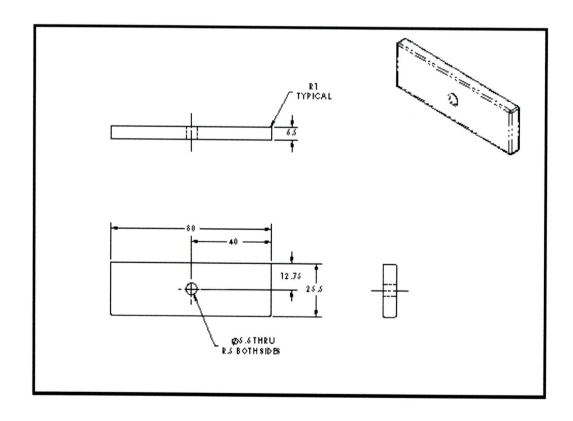

Handle:

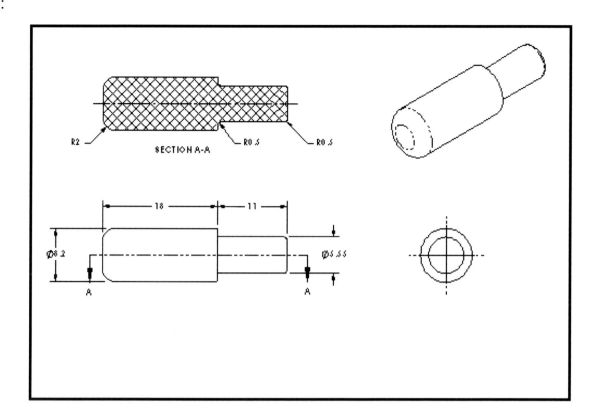

Assembly:

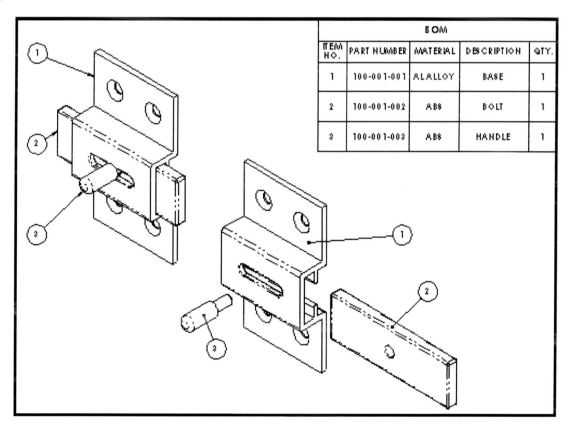

BOM				
ITEM NO.	PART NUMBER	MATERIAL	DESCRIPTION	QTY.
1	100-001-001	AL ALLOY	BASE	1
2	100-001-002	ABS	BOLT	1
3	100-001-003	ABS	HANDLE	1

Reflection Question

1. What do you think are the two most important reasons for always having a title block?

Quiz

1. Which of these items would you consider a standard part?

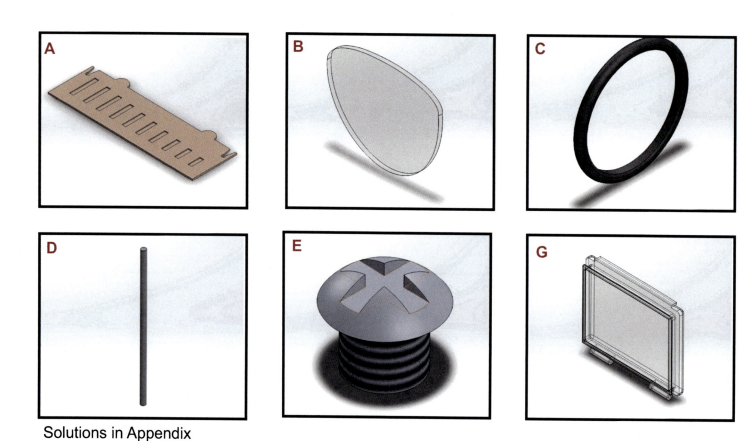

Solutions in Appendix

UNIT 6

DIMENSIONING STANDARDS

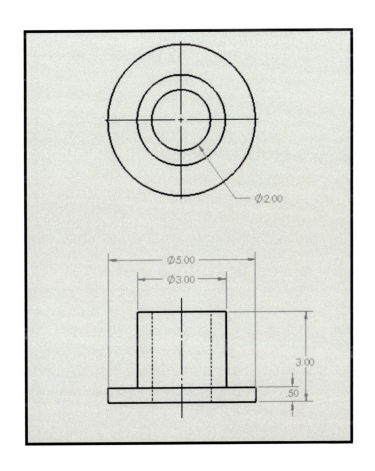

Introduction

While not the most glamorous technical graphics topic, dimensioning is crucial to communicate both size and location information for an object and its features. When you are making a product, you want it to be created exactly as designed, so dimensioning is employed to set appropriate constraints. A designer is able to communicate the design specs to a manufacturer because of dimensioning. This enables an object to be created and replicated again and again. Not only is this important for a single object, but also for ensuring multiple parts fit together correctly in an assembly. Chances are you have needed to buy a replacement case for one of your devices, such as a cell phone or tablet as in figure 6.1. Once you identify the correct model of case, installation is typically quite easy. This interchangability is a result of reproducing a design with proper dimensioning techniques.

Dimensioning is the language of technical graphics. Just like learning a foreign language, you will need to learn about the alphabet (line types), sentence structure (alignment and placement), special cases (like machined holes), and good grammar (good dimensioning practices). We will start with basic objects to highlight these topics. The good news is with practice interpreting and then creating dimensioned drawings, you will soon be able to interpret a wide variety of technical engineering documents. Technical drawing standards are the same across disciplines making this a very broadly applicable topic.

Figure 6.1 - Tablet Cases - Image courtesy of Hama at Wikimedia

While there are many standards outlined by professional organizations such as the American National Standards Institute (ANSI) and American Society of Mechanical Engineers (ASME), we will cover the most important rules and practices used in dimensioning objects in this unit. Rather than give examples of poor dimensioning practice, the examples in this unit highlight good dimensioning practices so you can refer back to them later when you are working through assignments or activities on your own.

Everyday Uses

While you may not be accustomed to reading dimensions from a technical drawing, you probably have had practice using dimensioning principles in your everyday life. For example, if you were looking at apartment or house plans, you may look at the room dimensions to see if the space would fit your furniture. Conversely, if you are looking to purchase a new couch, you may measure the couch first to see if it will fit in the space where you want to place it. Dimensions on a technical drawing are frequently a smaller scale (inch or millimeter), but the idea of using this information to communicate size and location ideas is very common.

Figure 6.2 depicts a hotel room layout that is compliant with the American Disabilities Act. In this case, the overall room dimensions are given as these are the minimum dimensions needed to fit generic hotel furniture while leaving the proper space for wheelchair mobility.

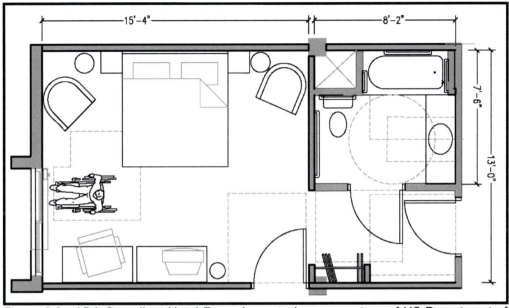

Figure 6.2 - ADA Compliant Hotel Room Layout - Image courtesy of US Department of Justice Civil Rights Division

Within the past decade, flat screens have replaced bulky, heavy, tube-based television models. In an electronics store, TVs are often arranged by their size with the dimension taken across the diagonal of the screen (see figure 6.3 The screen's dimension may be a deciding factor to fit into a certain space (like above a fireplace) rather than just the cost of the item. Beyond TVs, the size of large appliances (refrigerators, washer/dryers) is often a deciding factor for people when making a purchase. As a consumer, you use dimensioning principles to help you make decisions. Now we will discuss how to use dimensioning from a design standpoint to communicate technical information.

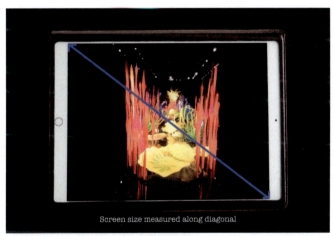

Figure 6.3 - Screen Size Example

With smartphones continuously evolving, dimensions are often used to highlight improvements over the previous model. For example, figure 6.4 shows how the screen size changes over various models of the iPhone (models 3, 5, and 7 displayed). As technology advances, it is possible to design better, larger screens that are potentially also thinner and lighter.

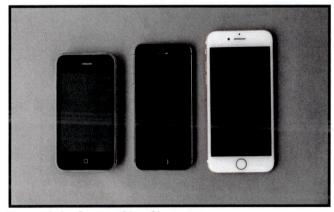

Figure 6.4 - Screen Size Changes

Shoe shopping is a another great example of needing to use dimensioning principles in everyday life. When looking for sneakers (see figure 6.5), you will commonly have your foot's length and width measured so you can select the best shoe for fit and comfort. The Brannock device used to measure feet has spaced graduated marks based on English units. Shoes are also a good example of an item that often requires unit conversion as shown in figure 6.6. You can find many conversion tables that show American shoe sizes versus European sizes. Often, you will see both of these sizes on shoe boxes or printed inside the shoe.

Figure 6.5 - Sneaker

Figure 6.6 - Sneaker Size

Basics

There are general rules for dimensioning that will keep your technical drawing looking neat and help you decide the best arrangement for placing dimensions on an object. There are often multiple correct ways to dimension any object, which can make the concept tough when students are beginning.

Typically, dimensioning is done on the top, front, and right side views of a orthographic multiview projection drawing, but not on an isometric pictorial projection. Basic dimensioning gives two main types of information: size and location. By size, we mean the overall size of an object or size of features of an object. By location, we mean where the features are located relative to the overall size of the object. Good practices for alignment, placement, unit notation, and overdimensioning will be discussed in this dimensioning unit.

For example, the schematic of the ratchet tool in figure 6.7 has the important dimensions labeled A through E. Dimensions that correspond to a single component like B, C, D, or E, indicate these sizes are important in the ratchet design giving necessary size dimensions. Dimension A shows the location of B relative to the edge of C making this a location dimension.

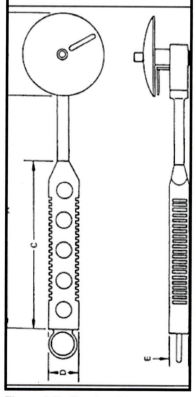
Figure 6.7 - Ratchet Dimensioning
- Image courtesy of NASA

Unidirectional Dimensioning

The most common dimensioning style is unidirectional, as shown in figure 6.8. Here the values are all read in the same orientation, which is typically parallel to the bottom edge of the drawing. In this unit, we will use unidirectional alignment in our dimensioning examples.

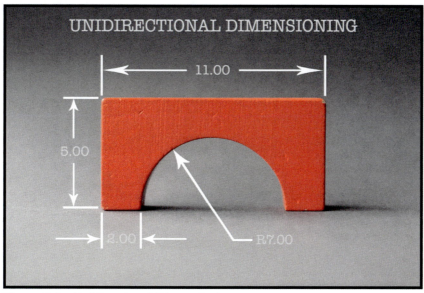

Figure 6.8 - Unidirectional Dimensioning

Placement

A number of products are on the market to help children develop fine motor skill coordination. For example, many toys involve the use of plastic nuts, bolts, and plates to copy simple designs or build own objects. For a toy like this to work as intended, the parts need to be interchangeable. Figure 6.9 is a simple base plate that we will use to examine the two main dimensioning placement methods.

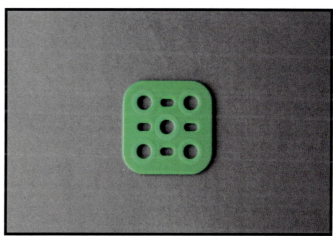

Figure 6.9 - Base Plate

The main question we need to ask in this case is what is more important, the position of the holes relative to the edge of the plate or the position of the holes relative to each other?

Let's first consider if the position of the holes relative to the edge of the plate is important. As shown in figure 6.10, we would use baseline dimensioning. The relative location of all of the holes would be made to the most important edge.

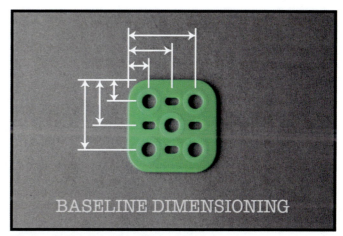

Figure 6.10 - Baseline Dimensioning

Now, let's consider if the position of the holes relative to each other is important. As we see in figure 6.11, we would use chain dimensioning. The dimensions between each hole would be given in this dimensioning placement method.

Mathematically, baseline and chain dimensioning are equivalent. What separates the methodologies is the design intent. If the distance to an edge is the most important design consideration, then use baseline dimensioning. If the distance between the features of the design is the most important for interchangeability, then use chain dimensioning.

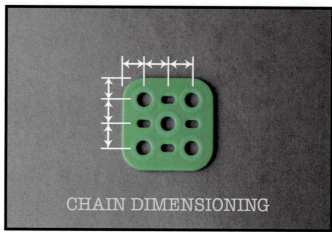

Figure 6.11 - Chain Dimensioning

Units

While many students in the United States are accustomed to using the English unit system, many international companies use metric units. To help distinguish between the two measurement systems, standards have been developed regarding how to display the numeric value of the dimension. Let's look at the measurement of an American Quarter (in English units) compared to a European Euro (in Metric units) in figures 6.12 and 6.13.

Figure 6.12 - American Quarter and European Euro

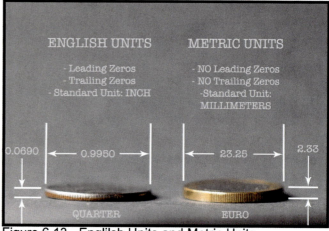

Figure 6.13 - Englilsh Units and Metric Units

Positive vs. Negative Cylinders

Now we will consider curved surfaces. A positive cylinder refers to a solid object with a single or double-curved surface. A negative cylinder is essentially a hole. It is important to be able to differentiate between positive and negative cylinders because they are dimensioned differently.

As an example, let's look at the top and front views of the glazed donut in figure 6.14 on the following page. Based on the top view, it is clearly round with a thru hole. The donut itself is a positive cylinder, and the hole is a negative cylinder. The positive cylinder looks like a rectangle in the front view shown in figure 6.14. When dimensioning, positive cylinders are dimensioned in the rectangular view. Negative cylinders are dimensioned in the circular view. While we will discuss annotation in a different unit, it is important to note that curved surfaces do require an additional symbol in front of the numeric dimension.

Figure 6.14 - Positive and Negative Cylinders

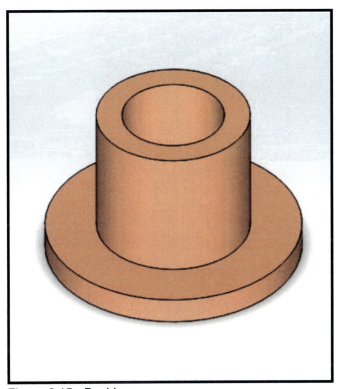

Figure 6.15 - Bushing

A bushing, like the one in figure 6.15 is an example of a standard part with both positive and negative cylinders. In this case, multiple positive cylinders are stacked on top of each other and the hole again goes through the entire object (thru hole).

How would we dimension an item like this? Let's look at the top and front view in figure 6.16. Remember that since this is a symmetrical object, we only need two multiview projection drawings to show the object. Also, note the use of hidden lines and centermark notation.

The negative cylinder (hole) is dimensioned in the top view where you see it as a circular view. The positive cylinders are dimensioned in the front view where they appear as a rectangle. Do you see the difference in the positioning? Another way to think of this is that you want to dimension positive cylinders in the view where you see the centerline, negative cylinders in the view where you want to see the centermark.

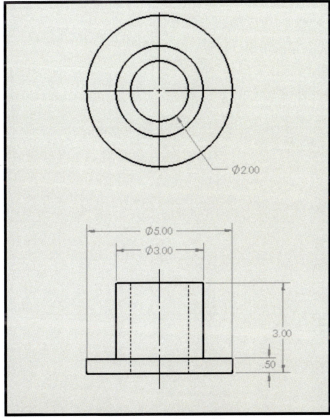

Figure 6.16 - Bushing Top and Front View

Figure 6.17 - Bushing

Tolerances

Tolerance is the amount a dimension can vary. Tolerances are important so parts fit together correctly in their assembled positions and the object functions as intended. While we will not discuss tolerance notation in detail, general tolerances are often noted in the title blocks for orthographic multiview drawings.

When tolerances are ignored, there can be serious consequences. If the tolerance for your door key is outside of the acceptable bounds, you will be locked out of your house (see figure 6.17). This is certainly an inconvenience, but these errors can be much more distressing. One example of a tolerance error is a bridge over a railroad that was designed 6 inches too short. The bridge was unintentionally built outside of the acceptable limits leading to the double stacked containers on trains not being able to pass underneath the newly, and costly, constructed bridge. Engineers had to spend about half a million dollars to redesign and raise the overpass.

Reflection Question

1. When have you used dimensions of an object to help you make a decision about a purchase?

UNIT 7

DIMENSIONING ANNOTATION

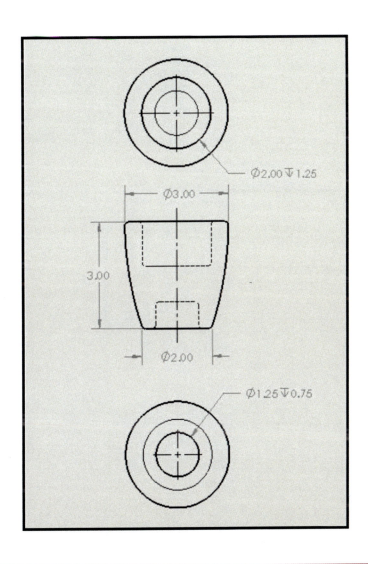

Introduction

Dimensioning is an important topic in technical graphics because it gives the information about how to take a 2-D object from a concept on paper to an actual product or prototype. Dimensions give the overall size information and tell you about the different features of a part so it can work in an assembly as intended. Dimensioning is a topic that is applicable in many different technical fields, and understanding the annotations will allow you to read and create a wide variety of documentation needed in the design process.

For example, the orthographic multiview of the space shuttle assembly image in figure 7.1 uses dimensioning principles to give information about the overall size in both metric and English units. While airplane hangers are obviously well-equipped to handle large aircraft, these overall dimensions are important when designing a space to permanently house the space shuttle. Based on the given dimensions, we can look at just how large this spacecraft is by simply looking at its footprint compared to the size of a regulation sized football field.

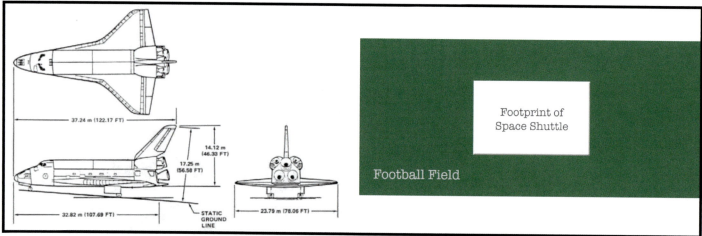

Figure 7.1 Shuttle Multiview and Size Comparison - Image courtesy of NASA

Knowing that size comparison, it may be easier to understand the designed space for the Enterprise at the Intrepid Sea, Air & Space Museum in New York as seen in figure 7.2.

Figure 7.2 - Shuttle Enterprise - Image courtesy of NASA

Everyday Examples

Having a set of annotation standards for dimensioning allows you to read a wide variety of technical documents and take the information on paper to an actual product. For example, architectural plans are generally required by municipalities for any type of new construction. These plans may not have been official or finished blueprints, but the idea of the structure design along with the associated sizes are clearly communicated. The image in figure 7.3 was hand-drawn in 1934 for a historical house in Troy, NY. Looking to the far left of the image, you can clearly see the aligned, chained dimensions that detail the height of each story of the building. This also clearly shows the block lettering expected in technical documentation.

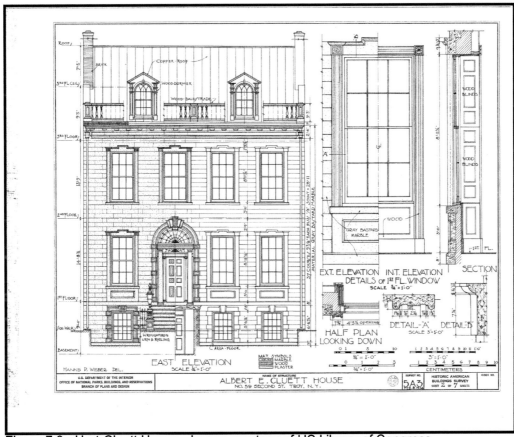

Figure 7.3 - Hart-Cluett House - Image courtesy of US Library of Congress

Dimensions are often used as a benchmark for compliance and determining design constraints. The Occupational Safety and Health Administration (OSHA) sets standards to help develop and maintain safe workspaces. In fact, companies and institutions may be audited by OSHA to make sure that they are in compliance with their practices and that their spaces are meeting requirements. The drawing in figure 7.4 shows height and width standards for various stair railings which are required in public spaces. This information is useful during the railing installation process and also meets requirements under the American Disabilities Act.

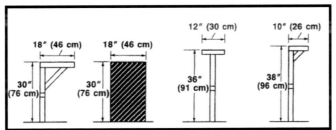

Figure 7.4 - Railing Standards - Image courtesy of OSHA

Issues with dimensioning can lead to some very expensive consequences. One of the best known examples is the mirror of the Hubble Telescope. When the instrument (see figure 7.6) was first launched into space, it returned blurry, unusable images and was initially deemed a disappointment. It turns out that during construction, a small washer was accidentally placed under the mirror during the cutting process shown in figure 7.5 and that changed the curve of the mirror just enough for the optics to not focus correctly. Fortunately, engineers were able to develop a solution to correct the dimensioning error (see image 7.7). The astronauts were able to install the Corrective Optics Space Telescope Axial Replacements (COSTAR) to make the Hubble Telescope functional allowing it to capture amazing images of the universe like the one in figure 7.8.

Figure 7.5 - Hubble Mirror - Image courtesy of NASA

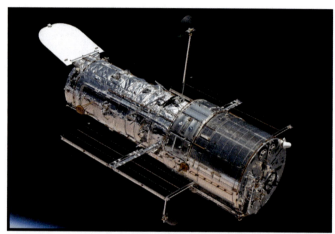
Figure 7.6 - Hubble Telescope - Image courtesy of NASA

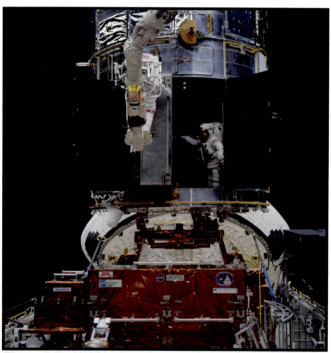
Figure 7.7 - Hubble Repair - Image courtesy of NASA

Figure 7.8 - Hubble Image Quality - Image courtesy of NASA

Dimensioning takes practice. While there are set standards to help guide you, rarely is there only one way to dimension any given object. It is important to have a firm grasp on good practices for dimension placement and annotation so you can clearly communicate the design idea. As you learn how to properly dimension an object, keep in mind the overarching theme of "clarity."

The drawing in figure 7.9 uses dimensioning annotation to show additional details about the dimensions.

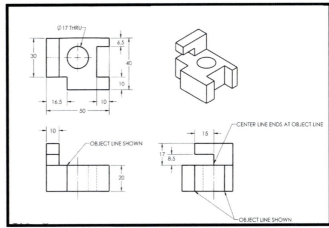

Figure 7.9 - Dimensioning Annotation

Line Types

Look at the line types in figure 7.10. Extension lines are drawn from important reference points and dimension lines with arrowheads are placed between those extension lines to give the size information. Notice how care is taken to stack the dimensions so they do not overlap and can be clearly read. This unit will discuss line types, dimensioning of arcs, and notation for machined holes.

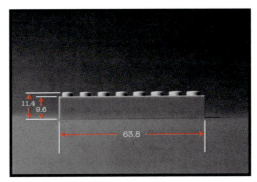

Figure 7.10 - Dimensioning

Let's start by deconstructing the different types of lines that are needed to make a dimension on a drawing. First, we have extension lines that stretch out from what you want to measure on an object, as seen in figure 7.11. These extension lines are typically thin and straight with a small gap between the object and the line itself. Extension lines may cross over each other, but it is good practice to avoid this when possible. A good rule of thumb is to not dimension to hidden lines, so extension lines really should pair up with a visible line in the view. The placement of extension lines dictates the placement of dimension lines, so it is good to pay extra attention when placing these extension lines in a dimensioned drawing (especially by hand).

Step 1 : Using a standard Lego brick, we will show how to place extension lines and dimension lines for the height and the width. When manually creating a dimensioned drawing, start with the placement of extension lines. For the Lego brick in figure 7.11, the extension lines are shown in green with the gap between the object and the line circled in blue. This gap between the object and the extension line is needed so it is clear where the bounds of object are shown by visible lines.

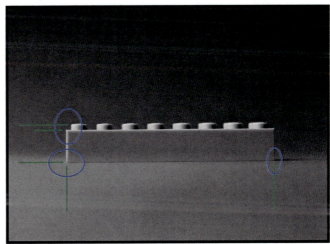

Figure 7.11 - Extension Lines

Step 2 : Next, with the extension lines places, dimension lines connect between extension lines with closed arrowheads at either ends (see figure 7.12). Dimension lines should not overlap with any other line types, which can be tricky as objects increase in complexity. Do not draw a dimension line directly to an object or to a centerline. Dimension lines should be paired with extension lines and they can be stacked as shown in red to the right.

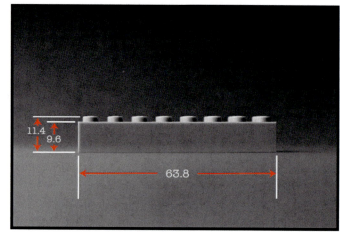

Figure 7.12 - Extension Lines with Arrows

Step 3 : Dimension lines are typically broken in the middle so a numeric value can be inserted, as in figure 7.13.

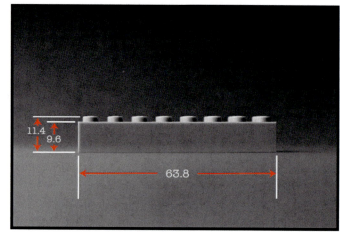

Figure 7.13 - Numerical Values

Figures 7.14 and 7.15 depict standard placements for the numeric values and for combining multiple dimensions.

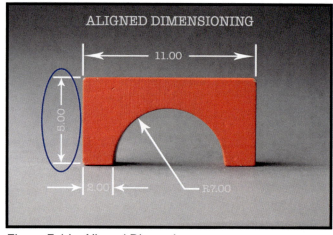

Figure 7.14 - Aligned Dimensions

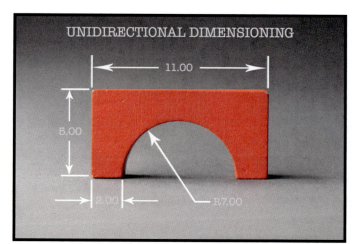

Figure 7.15 - Unidirectional Dimensions

What about dimensioning directly on an object? Whenever possible, dimensions should be placed outside the part. If there are truly no other better options for dimensioning, then you will see dimensions placed on a part, as demonstrated in figure 7.16.

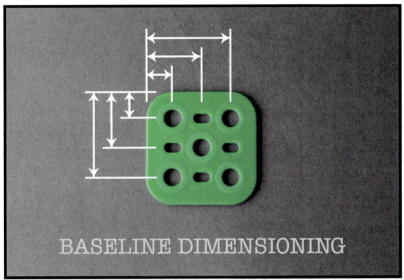

Figure 7.16 - Baseline Dimensioning

Arcs

Curved surfaces have their own dimensioning annotation rules to follow. Rather than use extension lines, the curved surface of interest is pointed at using a leaderline as in figure 7.17.

Use the symbol R in front of the numerical value for the radius of the arc.

Figure 7.17 - Curved Surface Dimensioning

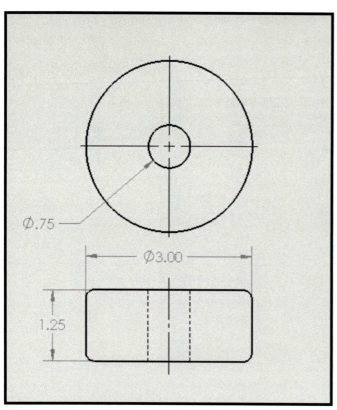

Figure 7.18 - Curved Surface Technical Drawing

The image in figure 7.18 is the same donut translated to a technical drawing. The dimensions are annotated the same as above with the diameter symbol used with circular features. In the technical drawing, we also include the correct annotation using line types like the centermark for the hole in the top view and corresponding centerline in the front view. Also, note the hidden lines are included in the technical drawing as compared to the images. Do you see how they align with each other?

Now let's take a look at an object with curved surfaces so we can examine how arcs are dimensioned. Arcs are curved lines or surfaces that are less than half of a full circle. In the case of the piece of ceramic art in figure 7.19, the sides have two major arcs. They are noted by using a leader line, but have the radius 'R' symbol in front of the numeric value. Contrast this notation with the diameter notation for the overall circular shape.

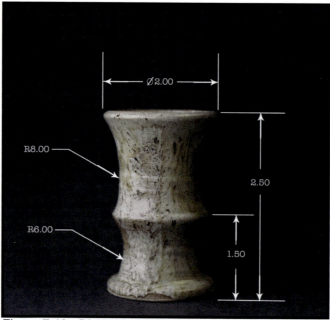

Figure 7.19 - Dimensioned Arcs

Machined Holes

While many examples in these units involve thru holes with a constant diameter, in reality, there are a wide variety of machined holes you may encounter such as blind, counterbored, and countersunk (see figure 7.20). This section serves to help you identify different types of holes and to introduce the specific dimensioning annotations to help define them.

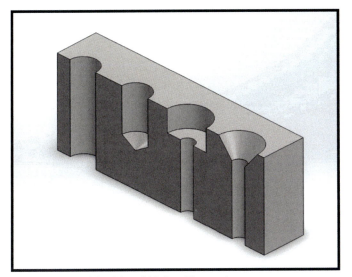

Figure 7.20 - Machined Holes Example

Blind Holes

Sometimes an object may have what appears to be a hole, but it does not go all the way through. This is referred to as a blind hole. The difference in the annotation is that the depth of the hole is given along with the dimension of the negative cylinder. The candle holder in figures 7.22 and 7.23 have a blind hole to hold a tea light candle, or it can be flipped over to hold a candle stick.

Often these machined holes are used to help with the connection or fastening of objects together. In figure 7.21, you can see the bolts sitting in their respective counterbored holes to help join a wooden beam to a bridge structure.

Figure 7.21 - Bridge - Image courtesy of OSHA

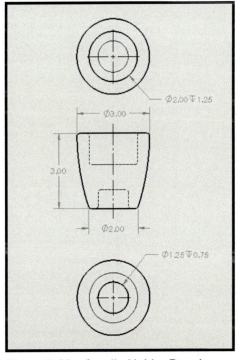

Figure 7.23 - Candle Holder Drawing

Figure 7.22- Candle Holder

Counterbores

If you have assembled furniture before, you may have seen a variety of fasteners with various bolthead shapes like those shown in figure 7.24. Often you do not want these fasteners to stick out past the edge of the object, so you screw them into a hole with a wider top so they are flush with the surface.

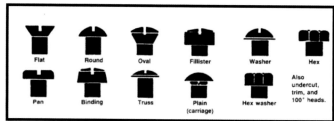

Figure 7.24 - Bolthead Shapes - Image courtesy of NASA

A counterbore is a type of machined hole that is frequently used. The counterbore differs from a regular hole with its wider top to accommodate the head of a fastener. The block in figure 7.25 has two thru holes in the upper corners and then a counterbore in the middle.

The image in figure 7.26 shows how a counterbore would look in a technical drawing. While we won't go into details of the annotation, counterbores require a short note to give all of the dimension information.

Figure 7.25 - Block with Thru Holes and Counterbore Hole

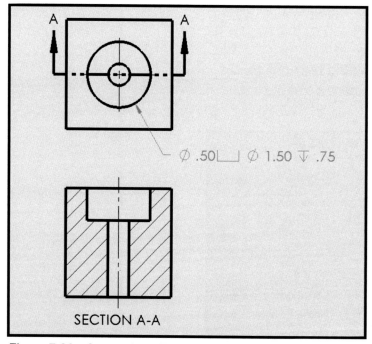

Figure 7.26 - Counterbore Technical Drawing

Countersinks

Countersink holes are another main type of machined holes in engineering design. Countersink holes tend to be used when greater precision is required. The main difference is that the top of the hole is angled. There are actually devices created to help measure the angles and depth of countersunk holes like those shown in the schematic in figure 7.27. Note the top of the countersink looks like a 'V'. The angle of the V may vary to accommodate different shaped fasteners.

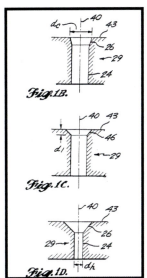

Figure 7.27 - Countersink Holes - Image courtesy of US Patent and Trademark Office

Countersunk holes are used often when it comes to hardware in a house. For example, wallplates for standard electrical outlets often have a single counterbore in the middle so when screwed into the wall the plate appears to lay flat (see figures 7.28 and 7.29). Cabinets may have countersunk holes drilled on the backside so when the hardware is installed the screw is flush (see figures 7.30 and 7.31).

Figure 7.28 - Wallplate

Figure 7.30 - Cabinet Countersink

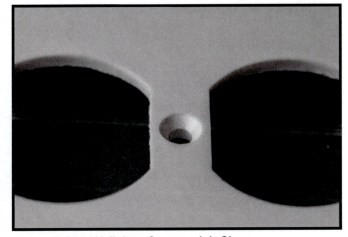

Figure 7.29 - Wallplate Countersink Close-up

Figure 7.31 - Cabinet Countersink Flush

Rounds and Chamfers

Paper is a material that lends itself to making boxes with sharp corners like we see on the shoebox in figure 7.32. The same can not be said for containers made out of glass. As seen in figure 7.33, the corners of the glass containers are rounded. These rounded corners are annotated differently and use the 'R' notation used with arcs. The rounded edge may also be referred to as a 'fillet'. Many times fillets are used to make edges or corners less sharp for safety or aesthetic reasons.

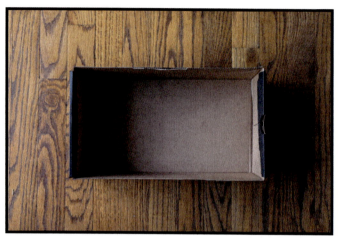
Figure 7.32 - Shoebox

Figure 7.33 - Glass Baking Dish

While not as common as rounds or fillets, chamfers are another type of detail you may find on objects. Rather than rounding an edge, chamfers take a small angled cut off of the edge to make it appear slightly inclined. Chamfers are often found in mats to mount images (see figure 7.34) and frames (see figure 7.35). They are also often used for aesthetic purposes like rounds and fillets.

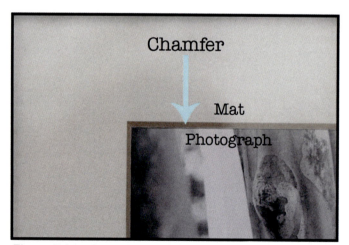
Figure 7.34 - Chamfer on Mat

Figure 7.35 - Chamfer on Frame

Reflection Question

1. While we looked at the width and the height of the Lego brick in figure 7.36, to fully dimension the object information needs to be included for the pegs on top. Are the pegs an example of positive or negative cylinders?

Quiz

Looking at the schematic of the Mercury 6 Rocket in figure 7.37, answer the following questions to practice interpreting a dimensioned drawing.

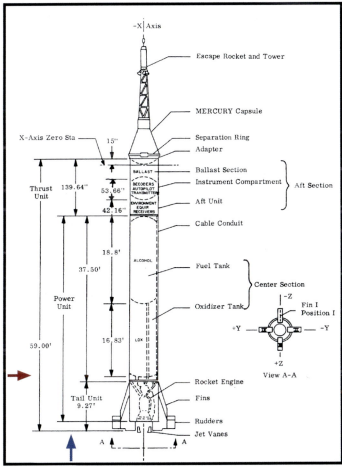

Figure 7.37 - Mercury 6 Rocket - Image courtesy of NASA

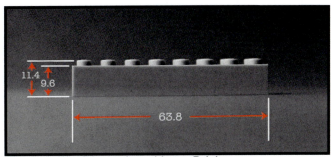

Figure 7.36 - Dimensioned Lego Brick

1. What is the height of the Thrust Unit? 15, 29, 42, or 59

2. Is this in English or metric units?

3. Would you classify this as chain dimensioning, baseline dimensioning, or a combination of both dimensioning methods?

4. Is the line pointed to with a vertical blue arrow an extension or a dimension line?

5. Is the line pointed to with a horizontal red arrow an extension or a dimension line?

6. Which line type uses a closed arrow - Extension or Dimension?

Solutions in Appendix

Activity

1. Knowing that dimensioning within an object should be avoided, sketch the side piece of the bird shelter in figure 7.38 and place the dimensions outside of the object. You do not need to include the numeric values, just draw the appropriate extension lines and then show where you would place the dimension. Your sketch should show six dimensions. Remember extension lines can be drawn inside an object for clarity, but avoid is placing the dimension inside the object.

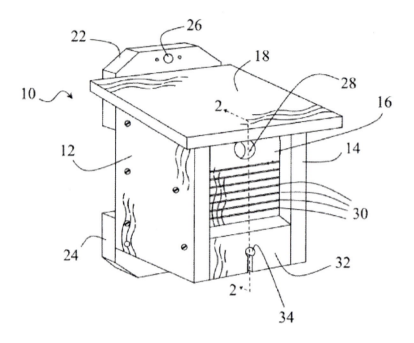

Figure 7.38 - Bird Shelter - Image courtesy of US Patent and Trademark Office

2. Figure 7.39 shows a truncated cone that will need three dimensions. Draw the extension and dimension lines. Assume the diameter of the top is 'A', the diameter of the bottom is 'B', and the height is 'C'.

Figure 7.39 - Truncated Cone

3. Place extension and dimension lines on the basic geometric shape with round features in figures 7.40. and 7.41

Figure 7.40 -Half Circle Block

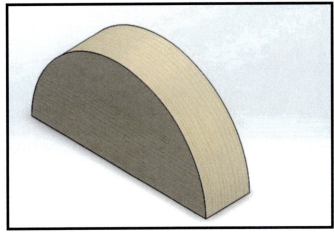

Figure 7.41 -Solid Model of Half Circle Block

4. Many fasteners consist of nuts and bolts. The exact size of the hexagon nut and thru hole in figure 7.42 are not known. Sketch the front and right side view of the object and then place your dimensions with generic labels.

Figure 7.42 -Solid Model of Half Circle Block

UNIT 8

ASSEMBLIES

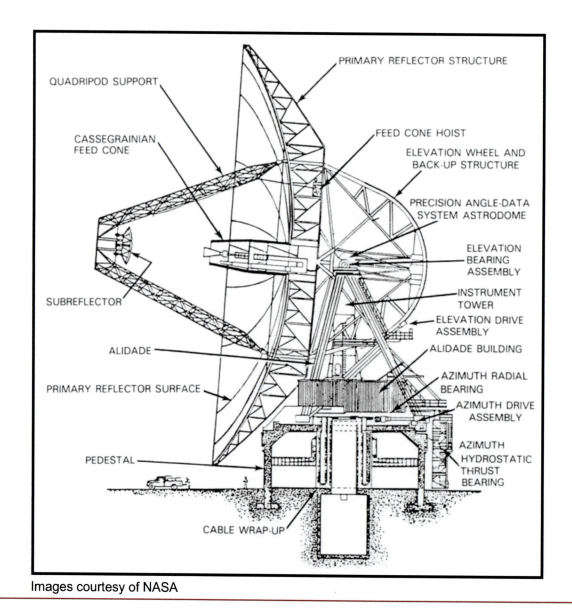

QUADRIPOD SUPPORT

PRIMARY REFLECTOR STRUCTURE

CASSEGRAINIAN FEED CONE

FEED CONE HOIST

ELEVATION WHEEL AND BACK-UP STRUCTURE

PRECISION ANGLE-DATA SYSTEM ASTRODOME

ELEVATION BEARING ASSEMBLY

SUBREFLECTOR

INSTRUMENT TOWER

ELEVATION DRIVE ASSEMBLY

ALIDADE

ALIDADE BUILDING

PRIMARY REFLECTOR SURFACE

AZIMUTH RADIAL BEARING

AZIMUTH DRIVE ASSEMBLY

PEDESTAL

AZIMUTH HYDROSTATIC THRUST BEARING

CABLE WRAP-UP

Images courtesy of NASA

Unit 8 - Assemblies

Introduction

The engineering design process often yields a product with many different parts that need to be put together in a certain way for the object work correctly. Assembly models and drawings are technical graphics methods used to communicate assembly instructions and indicate the correct operating position to the user. Three-dimensional pictorial CAD images and animations of assemblies are common ways for designers to show their product to technical and non-technical people.

The next generation Mars Rover in figure 8.1 is an example of an extremely complex assembly, but the concept is clearly visually communicated without accompanying technical details using a CAD-based assembly model. That's the beauty of assemblies–they not only look appealing, but also easily show the basic idea of incredibly complicated inventions.

Everyday Examples

Assembly views can be used in a variety of ways, including as an explanation for technical breakthroughs or failures, a proof-of-concept design before manufacturing, or a necessary support for legal documentation. One place you may have encountered assembly drawings are in infographics. For example, with the development of the Space Launch System by NASA, the new rocket was unveiled using a half-sectioned assembly model along with details and size comparisons so the public could more easily understand the scope of the project (see figure 8.2).

You have probably heard the saying, "A picture is worth a thousand words." In technical graphics, "An assembly is worth a thousand words."

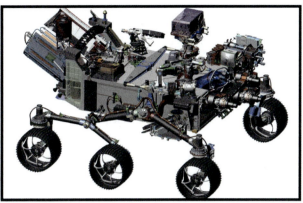

Figure - 8.1 - Mars Rover - Image courtesy of NASA

Assemblies play an important role in the iterative design process. Engineers and designers can test these prototype models in various software packages before constructing a physical prototype for real-world testing. In this unit, you will learn how to identify an assembly and how engineering geometry concepts are used to mate parts together. We will also discuss the difference between pictorial, sectioned, and exploded assemblies and advantages to each of these technical graphics documentation approaches.

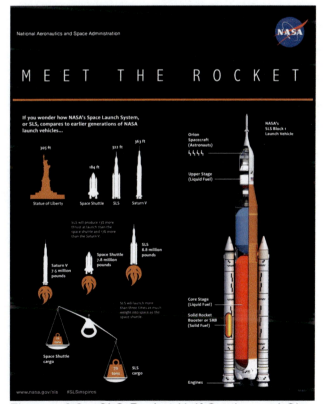

Figure - 8.2 - SLS Rocket Half-Section and Size Comparison - Image courtesy of NASA

Assemblies can be used to show proof-of-concept ideas. Before a product is actually produced, CAD-based assemblies are created and subjected to advanced modeling for optimization before actual production. With the increases in the precision of 3-D printing, engineers and designers are able to create custom parts more easily than ever before, allowing them to create complex designs with fewer overall parts. The exploded assembly in figure 8.3 is the concept for the first 3-D printed camera developed by NASA. This particular instrument is sized to fit in a 4x4" box so it can be incorporated into satellites. Ordinary cameras require approximately three times as many parts. Previously, they were constrained to traditional part creation techniques.

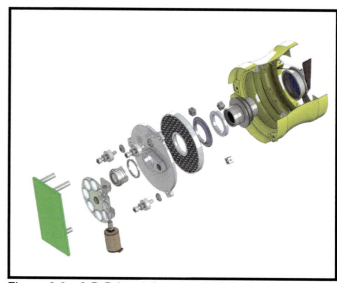

Figure 8.3 - 3-D Printed Camera CAD Assembly - Image courtesy of NASA

Using 3-D printed technology, engineers were able to drastically reduce the camera size while increasing resolution. The image in figure 8.4 shows the camera on the current Mars Rover. The image in figure 8.4 shows a 3-D printed camera for use in the next generation Mars Rovers. Assembly drawings can help explain major design changes between different iterations of a similar product.

Figure 8.4 - Older Mars Rover Camera - Image courtesy of NASA

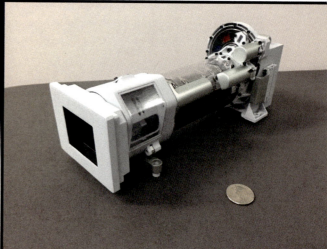

Figure 8.5 - Newer Mars Rover Camera - Image courtesy of NASA

Foundational Engineering Graphics: Principles and Applications

Assembly drawings are used extensively in patents. Many products include a drawing of the assembled object in combination with sectioned, exploded, or sub-assembly drawings to communicate the patented idea. For example, figure 8.6 displays an assembly drawing for a simple pitcher with a built-in water filter. To supplement this overall assembly drawing, the inventor also included a sectioned assembly drawing to better show the inner working pieces needed for the pitcher.

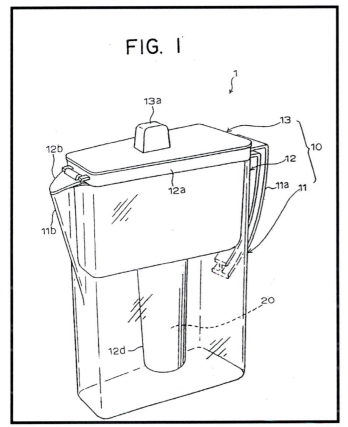
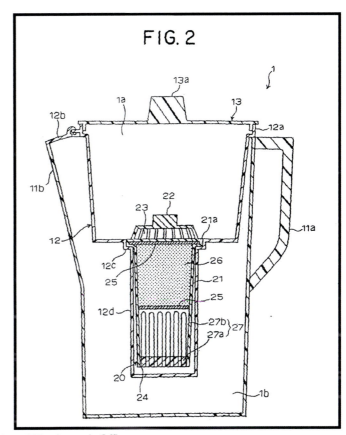

Figure 8.6 - Water Filter Pitcher - Images courtesy of US Patent and Trademark Office

Basics

We often show assemblies as the completed product (pictorial assembly). A few important variations of assemblies can be employed to better show how parts are put together. These include exploded assemblies (see figure 8.7) and sectioned assemblies.

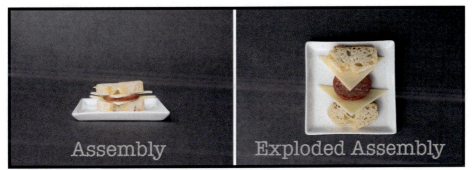

Figure 8.7 - Assembly and Exploded Assembly

What exactly constitutes an assembly? An assembly is any combination of two or more parts. An assembly can be very basic such as sunglasses (frame and lenses) to incredibly complicated (International Space Station in figures 8.8 and 8.9).

Assembly drawings are also a necessary component of a working drawing set. While drawing assemblies by hand is tedious work, today you may sketch a quick overview of the object and then quickly transition to making the various parts in a CAD-based software program. Creating assemblies in CAD-based software programs often employs ideas from engineering geometry and takes advantage of manipulating geometric relationships between objects.

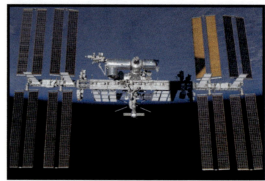

Figure 8.8 - International Space Station - Image Courtesy of NASA

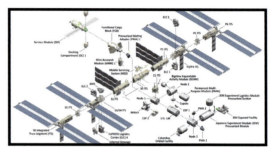

Figure 8.9 - International Space Station Assembly - Image courtesy of NASA

Pictorial Assembly

In the Pictorial Projection unit, we examined different types of 3-D rendering of single objects, one of which was a simple isometric view. A pictorial assembly builds on that concept and is an isometric of the entire assembly. This is the view you are most accustomed to seeing in everyday life on packaging, in advertisements, and with design pitches. Example are displayed in figures 8.10, 8.11, and 8.12.

Figure 8.10 - Press Pictorial

Figure 8.11 - Watch Pictorial

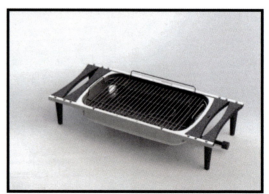

Figure 8.12 - Stove Pictorial

Sectioned Assembly

Since assemblies contain many parts, it is common for some parts to be hidden from view. Using a sectioned assembly is a great way to show hidden parts while also giving additional information about construction and correct operating positions. Section assemblies can be used with very complex structures such as the deep space antenna in figure 8.13 with parts labeled for clarity. In this technical drawing, note the creator included a truck and small person at the bottom to give a sense of size of the apparatus.

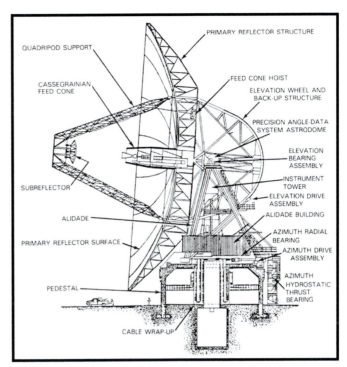

Figure 8.13 - Antenna - Image courtesy of NASA

Reflection Question

1. Now that you've seen some simple everyday assemblies, think about the objects that you currently have in your backpack or purse. Give three examples of assemblies that you use everyday. They can be simple or elaborate items.

Figure 8.14 shows samples of items you may have in your bag that would qualify as an assembly:

Figure 8.14 - Assembly Example

2. You are given the orthographic multiview of a custom designed water bottle. The bottle allows a user to store and drink two different types of drinks. From the top and front views as well as the pictorial in figures 8.15 and 8.16, how many parts does this object have? Do you think there may be any hidden parts?

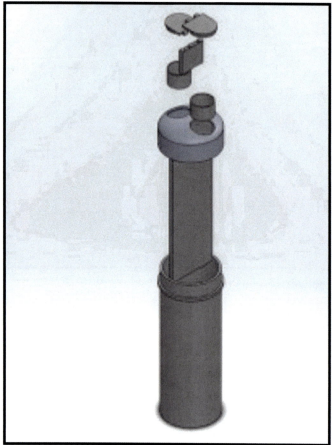

Figure 8.15 - Bottle Assembly

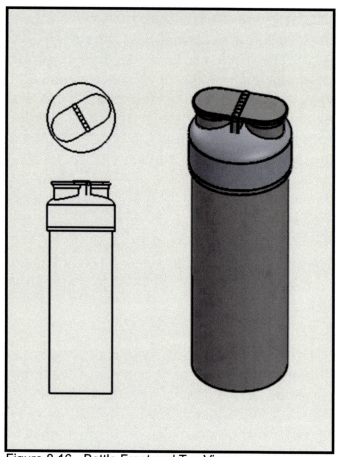

Figure 8.16 - Bottle Front and Top Views

Activity

1. Disassemble an available pen. How many parts does it have? Now, take the disassembled parts and align them as an exploded assembly. Would someone be able to assemble the pen based on your layout? Figure 8.17 offers a sample image to help you get started.

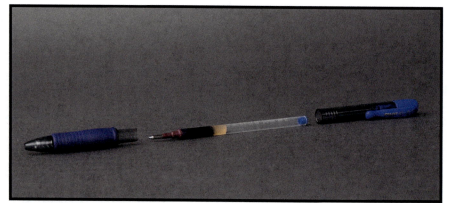

Figure 8.17 - Pen Assembly

---

1. For each of the following pairs of images, which item is an assembly?

1.

A

B

2.

A

B

3.

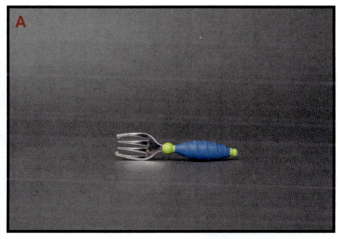
A

B

4.

A

B

Solutions in Appendix

UNIT 9

SECTION VIEWS

Images courtesy of NASA

Introduction

Section views are used extensively to show features of an object or an assembly that are not readily viewed from an initial observation (see figure 9.1). This method can be used with both simple and complex objects (see figure 9.2). Creating section views involves the use of a cutting plane that dictates what portion of the object you want to remove to reveal a more complex interior.

A full section view entails splitting an object in half can be helpful, but there are many more types of section views utilized in technical drawings to clearly describe the design. This segment will cover the different types of section views and corresponding technical vocabulary to help you determine which section view would best communicate important aspects of an object or assembly.

Everyday Uses

Have you ever visited an automotive dealership, a mattress store, or a window business? Chances are that you encountered a section view in a product advertisement as these views are also often used to show the quality of an object or to highlight differences between product levels or from a competitor.

For example, a mattress store may have a broken out section view sample for each different mattress so consumers can see how they are constructed differently. Examine the section view of the window in figure 9.3. This is used to show all the different parts of the window to a potential customer and then the differences in the panes or coatings can be explained in terms of energy efficiency.

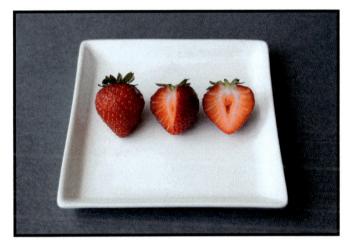

Figure 9.1 - Section View

Figure 9.2 - Complex Section View Image courtesy of Maximillian Schoenherr - Wikimedia

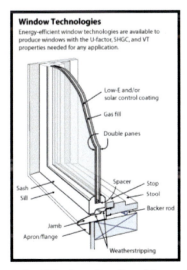

Figure 9.3 - Window Section View - Image courtesy of US Department of Energy

Figure 9.4 - Museum Section View Displays

While section views are used extensively in industry and technical drawing for documentation and communication purposes, they have a very practical application as a teaching tool to show the interior of complex objects in an informal manner. For example, figure 9.4 displays section views spotted at a science museum.

Section views can be used to show internal features of complex items that can not otherwise been seen from a pictorial projection assembly view. These types of technical drawings are important to show how different parts go together and used for formal documentation like that needed for a patent application. Figure 9.5 depicts a syringe apparatus consisting of a partial section to show the complex details of the tip more clearly. Section views help to better communicate complex design ideas and are often used in technical documentation.

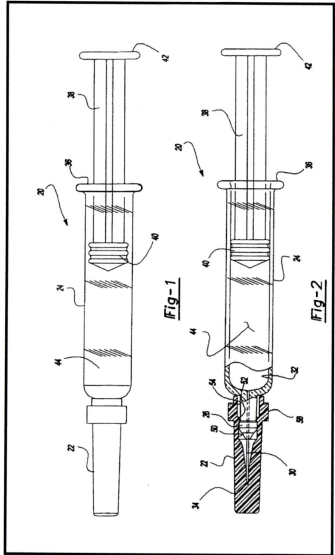

Figure 9.5 - Syringe Section ViewDrawing - Image Courtesy of United States Patent and Trademark Office

Basics

With section views, we are essentially using a cutting plane to slice an object or an assembly so we can see part of the interior. The cutting plane changes for different types of section views; it may be straight and well-defined or very jagged or somewhere in the middle. You can think of a cutting plane as literally the place where a cutting utensil like a knife or saw was used to slice an object. In the images in figure 9.6, the place where the pizza cutter slices the pizza is along the cutting plane. It can be sliced in a number of ways to make the desired sectional view.

Figure 9.6 - Pizza Cutting Plane

Some of the most common section views include removing half the object (full section) removing a quarter of the object (half section), or even a section view that looks like a literal bite was taken out of the object (broken out section). Let's take a look at some real world examples of the major types of sections views you are likely to come across in technical graphics and in engineering.

Full Section

A full section view means that you are removing half of the object. Slicing produce in half, such as the apple is figure 9.7, is an excellent example of a full section view.

Figure 9.7 - Full Section View

Half Section

A half section view means you are removing a quarter of an object. This type of view is ordinarily used when the object is symmetrical or if you only need to show a portion of a complex assembly. The cupcake images in figure 9.7 show the progression from the whole object to a half section to a full section.

Figure 9.7 - Half Section View

Assemblies

Half section views are most effective when used with symmetrical objects, whereas full section views are more applicable to a wide variety of objects. Full section views are more widely used than half section views. A telescope is a good example of a symmetrical, cylindrical object that would be shown well using a half section. The schematic is figure 9.8 demonstrates the basic configuration of the custom cylindrical mirrors of the Chandra X-ray Observatory telescope.

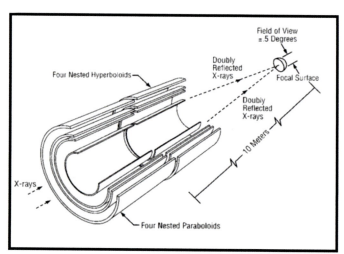

Figure 9.8 - Chandra X-ray Telescope Half Section View
- Image courtesy of NASA

Offset Section

Instead of making a straight single cut like in a full section, or two straight cuts in a half section, an offset section is a series of straight cuts to reveal interior features of an object. These are more conceptually difficult to imagine. An avid Lego builder may have seen offset sections when creating larger structures that open to reveal the important aspects inside the structure. The example is figure 9.9 shows a top and offset section view.

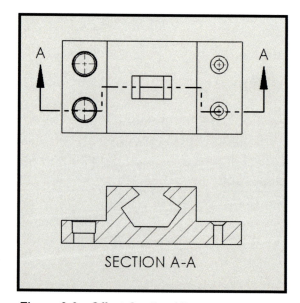

Figure 9.9 - Offset Section View

Foundational Engineering Graphics: Principles and Applications

Broken Out Section

A broken out section literally looks like someone has taken a bite out the object or assembly to give you a "peek" of the object's interior. These section view types are used when you are only interested in showing a specific portion of the inside of the object. This method is often used in illustrations to convey ideas to non-technical people. The illustration of the Gemini Spacecraft in figure 9.10 demonstrates a broken out section.

Figure 9.10 Gemini Spacraft Broken Out Ection View - Image courtesy of NASA

Revolved and Removed Sections

Revolved and removed sections are less common section views. A revolved section involves showing a section view in-line with the object (like the purple carrot in figure 9.11) and is often used to communicate more complicated cross-sections like in a support beam. Similarly, a removed section involves multiple cutting planes along the length of an object with the removed section views shown adjacent to the object (like the orange carrot in figure 9.11). Removed sections are useful for showing changes along the length of an object such as an airplane or spacecraft.

Figure 9.11 - Revolved and Removed Sections

Aligned Views

An aligned section view is essentially a special case of an offset section view. The cutting plane line is at an angle to show characteristics of non-aligned features. For example, let's look at the fidget spinner in figure 9.12. If we want to examine the holes and show that they are equally spaced from each other, we cannot do this with a straight cutting plane (like for a full section) or with a cutting plane bent at 90-degrees (like for a half section). In this case, the cutting plane is bent as seen in figure 9.13. The subsequent aligned sectional view shows the symmetry of the holes. Note this is not a true projection, so the section view appears taller than the actual object. This type of sectional view is often used with cylindrical objects with one plane of symmetry.

Figure 9.12 - Fidget Spinner

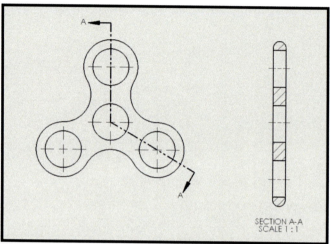

Figure 9.13 - Fidget Spinner Aligned Section View

Reflection Questions

1. Can you recall seeing an advertisement using a section view? What was it for? Has a sectioned view help convince you to buy a product?
2. Can you think of a display that you've seen that used section views to help show a more complex system?
3. When would you use a full section versus using a half section view of an object? Remember, a half section removes one quarter of the object and a full section removes half of the object as demonstrated with the strawberries in figure 9.14.

Figure 9.14 - Strawberry Section View

Quiz

1. Now that we've seen some examples of full, half, offset, and broken-out sections, let's practice identifying them. Hint: Identify the cutting plane line.

1. Bracket

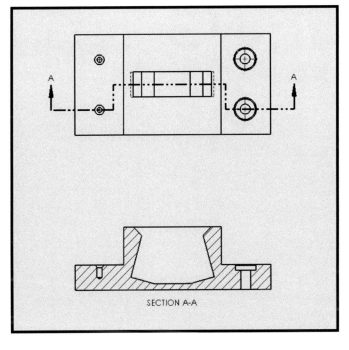

SECTION A-A

2. Water Bottle

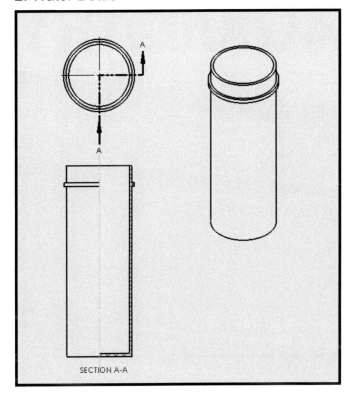

SECTION A-A

3. Piston

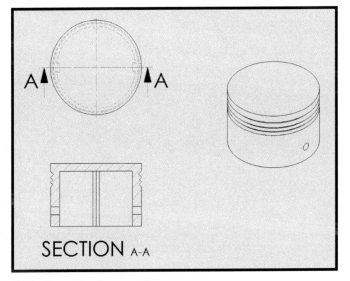

SECTION A-A

4. Connector

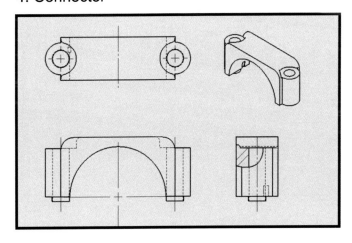

Solutions in Appendix

UNIT 10

AUXILIARY VIEWS

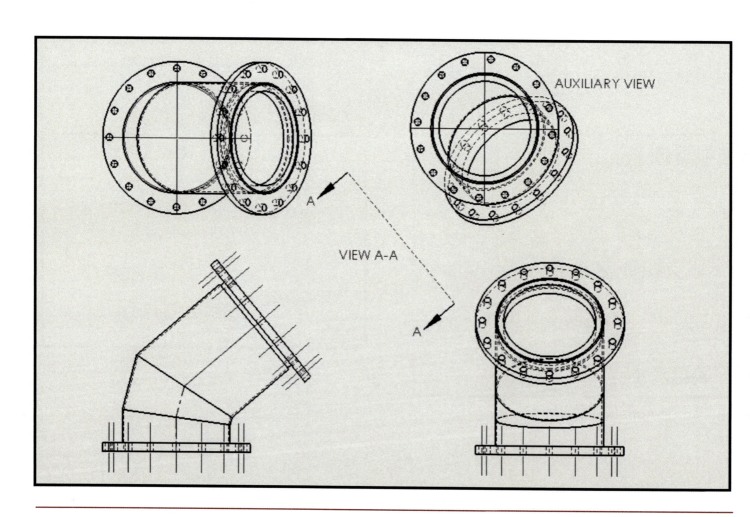

AUXILIARY VIEW

VIEW A-A

Foundational Engineering Graphics: Principles and Applications

Introduction

An auxiliary view is used to show the true size and shape of an inclined or oblique surface that can not be otherwise seen from any of the six principle views discussed in orthographic projection. You can think of an auxiliary view as a specialty view. It is only necessary for design clarity or dimensioning purposes for a non-normal surface. While this technique is used less often than a sectioned view, auxiliary views are very useful in the fields of surveying and construction.

For example, the American with Disabilities Act has a number of requirements regarding wheelchair-accessible buildings. A wheelchair ramp, like the ones in figures 10.1 and 10.2, is really just a large scale inclined surface. The true size of the ramp can be found using an auxiliary view.

Figure 10.1 - Wheelchair Ramp

Figure 10.2 - Additional Wheelchair Ramp

Creating auxiliary views by hand can be a laborious task, but there are methods to speed up the process. With the advent of CAD-software programs, auxiliary views are easily generated, which has increased the use of this specialty orthographic view. This unit will help you identify situations when an auxiliary view may be useful by showing the relationship of an auxiliary view to the principle views in accordance with Glass Box Theory.

Everyday Uses

An auxiliary view is a view from any image plane other than the frontal, horizontal, or profile. It is used to show the true size and shape of an inclined or oblique surface. The example in figure 10.3 shows an auxiliary view of a roof section. The blue section of the roof in the top view is foreshortened. It is shown in true shape and size in the orange auxiliary view.

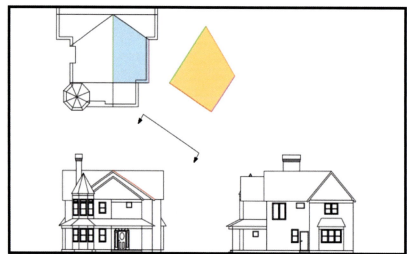

Figure 10.3 - Auxiliary View of Roof Section

Have you ever seen a house get a new roof before? You may have noticed that roofing companies typically do not have much shingle waste as they are able to calculate fairly precisely the number needed to cover the inclined surface of a roof (see figure 10.4). While there is computer software available to help with the materials required calculations these days, the code to do so is based on an understanding of auxiliary projections to find the true area size of all of the inclined surfaces that create a roof.

Figure 10.4 - Installing Shingles

Basics

We have established that auxiliary views are useful for describing the true size and shape of surfaces that would appear foreshortened in the principle views. Auxiliary views are useful for determining true surface size, reverse construction (like determining number of shingles on a roof), finding the true length of a line (think about a property line), as well as examining point and edge views, if needed.

While auxiliary views can be easily created in any CAD software, the main way to create an auxiliary view manually is the fold-line method. We won't go into all of the details here for constructing an auxiliary view by hand, but it is important for you to understand how an auxiliary view relates to the other principle views and be able to identify an auxiliary view in a technical drawing.

Going back to multiview orthographic projection, remember that the true size and shape of an inclined surface cannot be seen from one of the three principal views in an orthographic projection. Figure 10.5 shows a front view of a ramp. We can see the true size and length of the incline from the front view since we are looking at the edge view. If we shift our perspective to more of a left side view as in figure 10.6, the incline appears foreshortened which means that the incline appears shorter than it actually is in real life.

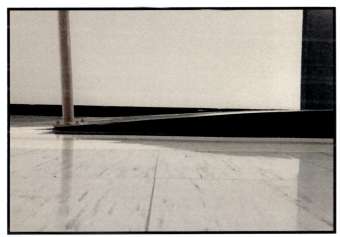

Figure 10.5 - Ramp Front View

Figure 10.6 - Ramp Auxiliary View

Fold Line Method

Recall the glass box model we used for orthographic projection and construction of multiview drawings. Remember how we could essentially "unfold" the box to get the six principal views? The fold-line method builds on this example. Think of adding in an extra panel that shows the auxiliary view of the inclined surface in true size and shape.

Full, Partial, and Half Auxiliary Views

Now that you can identify a surface that can be shown in an auxiliary view, let's look at these auxiliary views in the context of technical drawings. We will use an elbow pipe piece with flanges on both ends (see figure 10.7) to demonstrate the difference between a full, partial, and half auxiliary view. While many elbow pipe pieces are bent at 90 degrees, these pieces frequently need to be customized. As you can see from the orthographic pictorial of the flange, the bottom flange is aligned with the principal bottom view. The top flange however does not align with the right side view but rather is aligned with an inclined plane. Let's take a look how this piece can be documented using auxiliary views.

Figure 10.7 - Elbow Pipe

Full Auxiliary View

In a full auxiliary view, you show the entire object with all of the hidden lines and object features just like you would for any of the principle views (see figure 10.8). This can lead to the auxiliary view looking extraordinarily messy and difficult to correctly interpret. In this case, the auxiliary view is in the upper right corner. You see the flange in true size and shape but all of the hidden lines are distracting from the face that you are actually interested in. The hidden lines also make it difficult to add additional annotation such as dimensions to the full auxiliary view.

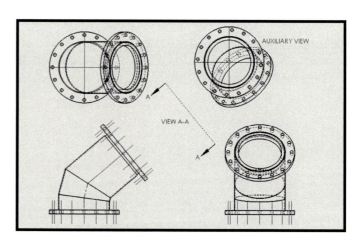
Figure 10.8 - Elbow Pipe Full Auxiliary View

Partial Auxiliary View

While in multiview drawings it was important to show hidden lines in each of the views, auxiliary views are treated differently. It is acceptable in practice to show a partial auxiliary view in place of a full auxiliary. In a partial auxiliary view, it is normal practice to omit hidden lines and show only the inclined surface of interest, rather than the whole object. Partial auxiliary views show only the true size and shape of the inclined surface making interpretation and proper annotation (like dimensioning) significantly easier.

In the case of the flange piece, a partial auxiliary view shows just the circular flange part (see figure 10.9). This view permits the easy dimensioning of this face. Do you see the difference between the full auxiliary and partial auxiliary views? Partial auxiliary views are preferred with more complex parts and often are used in place of full auxiliary views for clarity purposes.

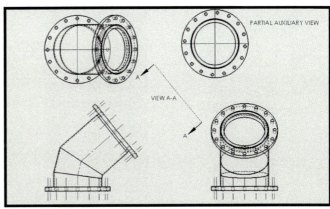
Figure 10.9 - Elbow Pipe Partial Auxiliary View

Half Auxiliary View

A half auxiliary view (see figure 10.10) is a subset of a partial auxiliary view that is used primarily with symmetrical objects. Rather than show the entire face of the flange, you would only show half as you can still give all of the information about the object using half of the face.

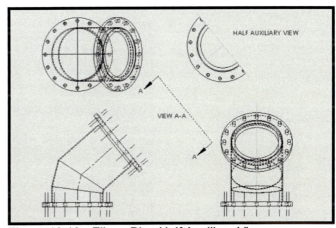
Figure 10.10 - Elbow Pipe Half Auxiliary View

Reflection Question

1. Auxiliary views have many practical applications in large scale construction projects as you have seen above. Can you think of an inclined object that may have used auxiliary view principles during the construction phase (design planning or materials purchasing planning)?

Activity

1. Here are some basic items with an inclined surface. Based on the isometric view, identify the top, front, right side, and auxiliary views.

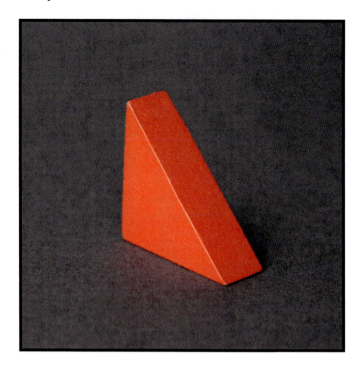

Engineering Graphics in Context

The text you have explored has provided the principles of engineering graphics that you can apply to numerous fields of study. Knowledge from this text enables you to communicate ideas to people in all types of industries all over the world. While languages may be different, you are able to communicate ideas through technical drawings detailing shape, form, various perspectives, details of materials, and dimensions. Careers that directly use engineering graphics include project managers, engineers, engineering and technology educators, product researchers and developers, architects, and designers just to name a few. Even if you do not plan on pursuing a career directly using engineering graphics, the skills branching from spatial visualization gained through engineering graphics are vital for careers that incorporate problem-solving, planning, working with data, and communicating details of designs.

When bringing your ideas to life, sketching and technical drawing are vital components of the design process. You begin by thinking of a design and then sketching your idea. This allows you to communicate to an audience through visuals what ideas you have conceived. From there you can take your sketch and redesign it as necessary adding in additional detail. Initially, your brainstormed ideas will start off as hand-drawn sketches. Ultimately, as your idea evolves you will move from hand-drawn sketches and technical drawings to the use of Computer-Aided Design (CAD).

After gaining foundational knowledge and experience with engineering graphics, you can advance by further exploring how to apply your knowledge to CAD software. CAD is the application of engineering graphic standards to computer software that allows you to digitally draw your design, modify it as needed, and even analyze your design through digital simulations. There is a variety of CAD software currently available for levels ranging from novice to expert. Once you have gone through the cycle of the design process and are ready to move forward with bringing your ideas into physical form, engineering graphics coupled with CAD will take you even further.

From technical drawings and CAD renderings, your designs can be used as plans to build your product. This is how the majority of the objects we interact with every day are developed. The majority of items used daily started out as a sketch, then a technical drawing, and then a physical product following the details of the technical drawing. With current digital technology and the use of CAD, people can use digital manufacturing where a machine controlled by computers can take a CAD file and develop a physical product. Examples of this include, but are not limited to, 3D printers, laser cutters, Computer Numerical Controlled (CNC) machines, and more.

Engineering graphics enable us to bring our ideas into reality by establishing a way to communicate the details of designs. Communication is a key skill to sucess in the workplace. Technology throughout the world continues to grow and it is greatly dependent on the ability to present ideas clearly from multiple perspectives. As you take this knowledge further into any field you choose, remember that there are tools in place to help you solve problems and to communicate with people on a global scale.

Unit 1 - Sketching

Reflection

2. Scissors

Quiz

1. C
2. B
3. A
4. C
5. B
6. A
7. A and B

Unit 2 - Engineering Geometry

Quiz

1. A
2. B
3. C

Unit 3 - Orthographic Projection

Quiz 1

1. Correct
2. Incorrect - Missing center line and center mark
3. Incorrect - Missing hidden lines and right side centerline.

Quiz 2

1. B
2. A
3. A
4. B

Unit 4 - Pictorial Projections

Activity

1.

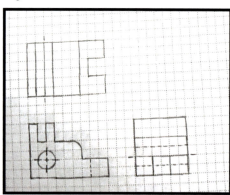

Unit 4 - Pictorial Projections

Activity

3.

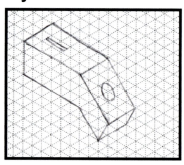

Unit 5 - Working Drawings

Reflection

Easy referencing or indexing at any point in the design process and archival purposes.

Quiz

C and E

Unit 7 - Dimensioning Annotation

Reflection

Positive

Quiz

1. 59

2. English The unit marks for foot and inches are included with the numeric values of the dimension. For example, the 'Thrust Unit' is noted as 59.00' (59 feet). The 'Aft Section' comprised of the Adapter, Ballast Section, Instrument Compartment, and Aft Unit is 139.64" (139.64 inches).

3. Both Remember that baseline dimensioning refers to dimensioning from an edge and chain dimensioning refers to dimensioning to items in order. In the case of the rocket, there are a parallel series of chain dimensions that all start with the uppermost extension line. So in this instance, the case can be made that a combination of baseline and chain dimensioning are present in the drawing.

4. Extension Line
5. Dimension Line
6. Dimension

Unit 7 - Dimensioning Annotation

Activity 2

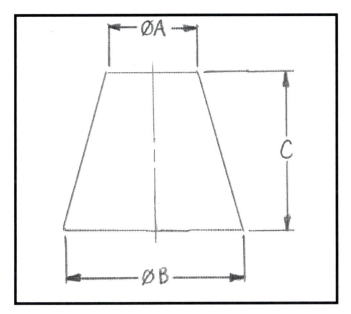

Activity 3

Since this object is a half circle, the height of the object is the radius.

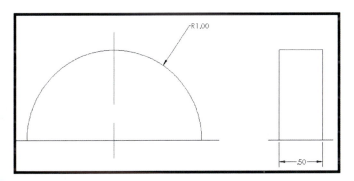

Activity 4

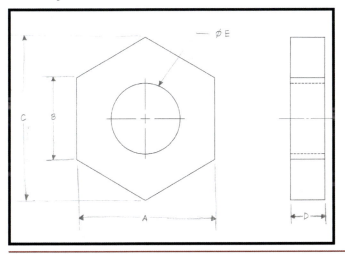

Unit 8 - Assemblies

Quiz
1. A
2. B
3. A
4. B

Unit 9 - Section Views

Quiz
1. Offset
2. Half
3. Full
4. Broken Out

Activity 1

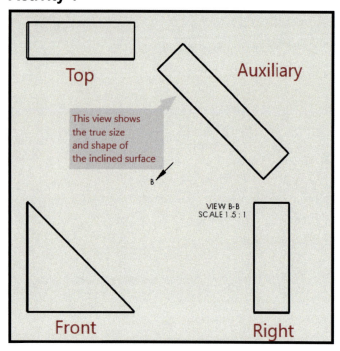

This view shows the true size and shape of the inclined surface

B

Top

Auxiliary

VIEW B-B
SCALE 1.5 : 1

Front

Right

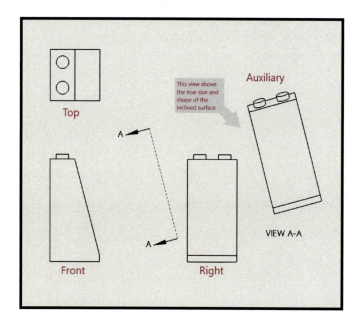

This view shows the true size and shape of the inclined surface

Top

Auxiliary

A

A

Front

Right

VIEW A-A

Reflection

1. Sample answers:
   - Railing length
   - Amount of carpet to cover stairs
   - Brinks for an incline
   - Size of a window for a skylight
   - Amount of tarp material to cover a patio (solar shade)

Appendix B - Paper Template Grid Paper

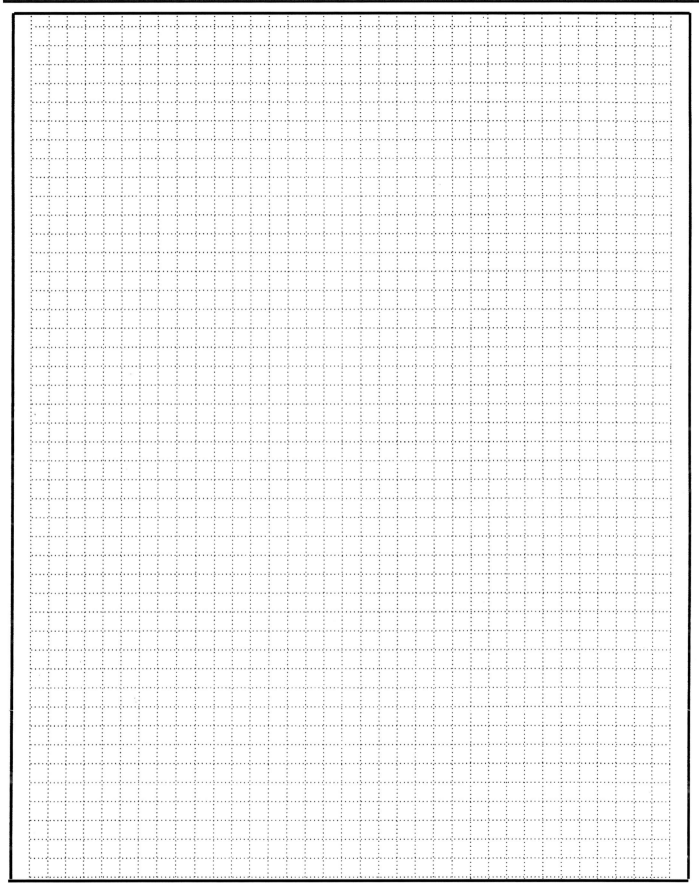

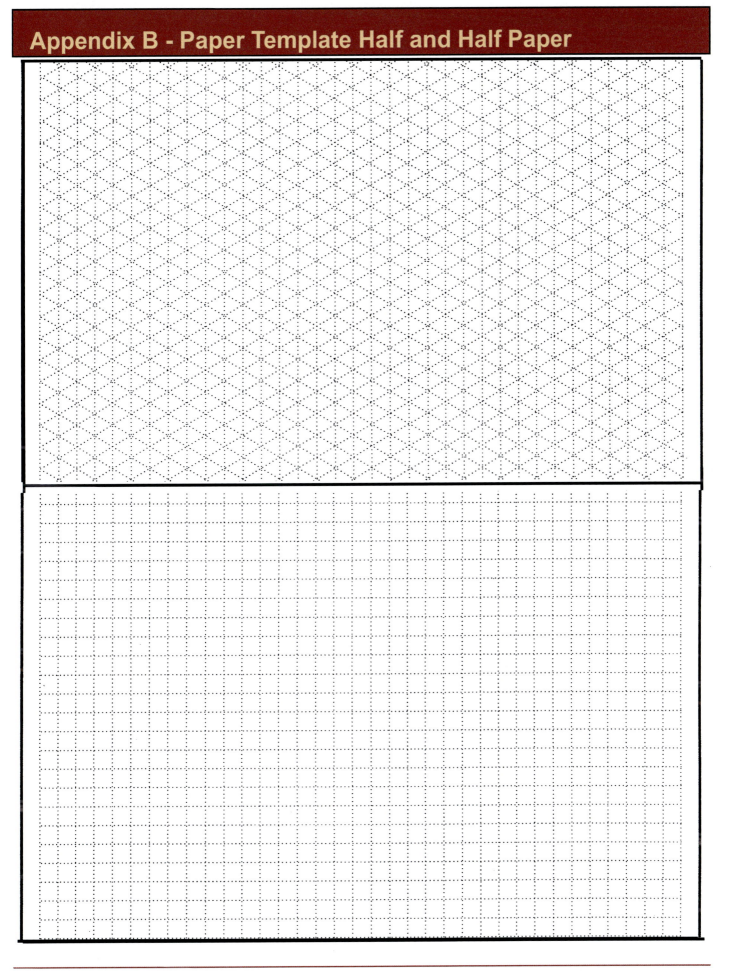

Made in the USA
Columbia, SC
29 June 2021

41166825R00066